DESIGN OF STRUCTURES AGAINST FIRE

Proceedings of the International Conference on Design of Structures Against Fire, held at Aston University, Birmingham, UK on 15th and 16th April 1986.

DESIGN OF STRUCTURES AGAINST FIRE

edited by

R. D. ANCHOR

Visiting Senior Lecturer, Department of Civil Engineering and Construction, Aston University, Birmingham, UK, and Senior Partner of R. D. Anchor Consultants, Birmingham, UK

H. L. MALHOTRA

Honorary Visiting Research Fellow, Department of Civil Engineering and Construction, Aston University, Birmingham, UK and Agniconsult, International Fire Protection Consultancy, Radlett, Herts., UK

and

J. A. PURKISS

Lecturer, Department of Civil Engineering and Construction, Aston University, Birmingham, UK

ELSEVIER APPLIED SCIENCE PUBLISHERS
LONDON and NEW YORK

ELSEVIER APPLIED SCIENCE PUBLISHERS LTD
Crown House, Linton Road, Barking, Essex IG11 8JU, England

Sole Distributor in the USA and Canada
ELSEVIER SCIENCE PUBLISHING CO., INC.
52 Vanderbilt Avenue, New York, NY 10017, USA

WITH 8 TABLES AND 136 ILLUSTRATIONS

© ELSEVIER APPLIED SCIENCE PUBLISHERS LTD 1986

(except for Papers 6 and 7)

© GOVERNMENT OF CANADA—Paper 7

British Library Cataloguing in Publication Data

International Conference on Design of Structures
Against Fire (*1986: Aston University*)
Design of structures against fire.
1. Buildings, Fireproof. 2. Fire prevention.
I. Title. II. Anchor, R. D. III. Malhotra, H. L.
IV. Purkiss, J. A.
693.8'2   TH1065

ISBN 1-85166-012-7

Library of Congress CIP data applied for

The selection and presentation of material and the opinions expressed in this publication
are the sole responsibility of the authors concerned.

Special regulations for readers in the USA

Printed in Great Britain by Galliard (Printers) Ltd, Great Yarmouth

Preface

Over the past two decades a considerable amount of effort has been directed towards producing what have been described as rational methods of design of structures to resist the effects of fire. It, therefore, also has become apparent that traditional methods of fire design using 'deemed to satisfy' clauses in Statutory Regulations or Design Codes of Practice are inadequate due mainly to their conservative and restrictive nature. Equally it has become recognized that the standard furnace test bears little or no relation to the actual structure behaviour in a real fire. Thus it has become necessary to propose new methods, and the approach known as Fire Engineering has been evolved.

In order to be able to produce calculation methods for fire design, it has been necessary to undertake research on the behaviour of structural materials at elevated temperatures. The current position here is exemplified by the recent reports on steel and concrete published by the RILEM Committee No 44-PHT under the chairmanship of Bill Malhotra. Calculation methods have progressed in two directions: those involving finite element techniques, including the solution of the heat transfer equations, and those involving simple numerical or graphical techniques. It is inevitable that whilst research will utilize the more complex methods of approach, the design engineer will need a simpler (but not necessarily simplistic) approach. At one end of the spectrum Fire Engineering may become a specialist discipline when dealing with large complex structures and unusual fire exposure conditions, but at the other, when dealing with normal structural problems, it should become assimilated into the normal design process.

The papers contained within this volume are concerned with evaluating the current state of Fire Engineering Design, and also with the ground work that will form the basis for an increased usage of such design methods in the future. In order to do this, five areas of importance were identified. These are dealt with individually:

1 *Fire Protection.* In this section the basic need for Fire Protection and the modelling of fire response is dealt with.
2 *Material Behaviour.* This section deals with modelling the behaviour of basic construction materials at elevated temperatures and their use in calculation methods.
3 *Design Concepts.* This section deals with the philosophy behind fire design—namely an extension of limit state methods—including computer applications.
4 *Design Implementation.* This considers the different approaches required for design in each of the common structural materials.
5 *Post-Fire.* The last section considers what happens after a fire and the measures needed to reinstate a structure.

It is anticipated that within the next decade, increasing advances will be made and that these papers will go some way to stimulating these advances.

The Editors wish to place on record their thanks to all the authors for having produced papers and for having been willing to share their knowledge with others. The Editors also wish to thank the Department of Civil Engineering and Construction, Aston University for their support during this venture.

R. D. ANCHOR
H. L. MALHOTRA
J. A. PURKISS

Contents

ix

Section 5: Post-Fire

A SURVEY OF FIRE PROTECTION DEVELOPMENTS FOR BUILDINGS

MALHOTRA, H L

AGNICONSULT
International Fire Protection Consultancy
44 Goodyers Avenue, RADLETT, Herts, UK.

ABSTRACT

Fire protection requirements date back to middle ages but only over the last 100 years or so has some pronounced progress been made in the technical content. Whilst the fire protection needs are understood their specification is still far from rationally based. More progress has been made in the field of fire resistance and fire severity estimation, but less so in the inter-actions which exist amongst different measures. For a rational system fire engineering principles need to be applied to the whole field of fire protection for buildings.

INTRODUCTION

Concern with fires in buildings can be traced back to the fire tragedies which have occurred in the past and devastated cities and towns. The burning of Rome in AD 64, the Great Fire of London in 1666, the fire in Hamburg in 1891 and the destruction of Baltimore in 1904 are only a few of the examples. In each case there was a follow up particularly with the objective of improving constructional standards for buildings, their separation from each other and fire fighting facilities. Despite these rules serious fires have continued to occur, perhaps with less frequency and with only restricted spread to other buildings. Concentration of low rise buildings with narrow streets has been replaced by high rise or high density occupancies each housing a few thousand people.

The objectives of fire protection in buildings can be summarised as:

1) safety of occupants, i.e. life safety,
2) minimizing fire damage, i.e. property protection, and
3) prevention of conflagrations.

Life safety is the declared objective of most governmental or local authority control although some of the requirements are set at a level where property protection is also assured to a certain extent. Occupants are put at risk at the start of a fire and it is often suggested that they move to a place of safety, inside or outside the building, in less than 30 minutes, hence requirements in excess of this period could have property protection implications. Whilst this assumption may be true for low rise, low density occupancies, in larger buildings it is often necessary to assure the stability of the structure for the whole process of fire development and to protect the adjacent buildings for as long as the fire is likely to last. This is particularly the case with high rise structures and complex assembly type occupancies such as shopping malls.

Property protection was at one time considered to be of significant interest to the community and fire regulations made no distinction between requirements for life safety or property protection. However, when the insurance companies came into being and initially had their own fire fighting equipment, property protection became a matter of business agreement between an owner and his insurance company. The fire brigades became the responsibility of local authorities in the eighteenth century and fire insurance developed new approaches to property protection. Isolation of risks by substantial fire barriers and the use of sprinkler installations has become the mainstay of insurance philosophy.

The current fire protection system in most countries is such that the central or local authorities make rules and regulations to ensure the safety of life and prevent the spread of fire from building to building and the property protection is left entirely to an insurance arrangement. Because some of the current regulatory requirements are a reflection of the past practices the measures have some relevance for property protection as well.

COMPONENTS OF FIRE PROTECTION

Comprehensive fire protection requirements need to consider a number of aspects from fire prevention to fire safety management. At least 10 distinct needs can be identified:

1) Prevention of fires,
2) Detection of fires,
3) Delaying fire growth,
4) Controlling movement of smoke,
5) Providing means of escape,
6) Restricting fire spread,
7) Preventing collapse of the structure,
8) Controlling fire,
9) Fire fighting, and
10) Fire safety management.

The first and last are usually not included in any building control system and are taken care of by some subsidiary system on the use of energy producing or using appliances and licencing requirements for the premises. Such measures also have no direct relevance to the construction of a building. The other measures can be divided into two categories, passive and active. Passive measures are those which are provided as an in-built feature of the building and are operative at all times, whereas the active measures may be provided during or after construction but they become operative only on the occurence of a fire. Eight of the measures listed above will be categorised as passive or active as below;

Passive	Active
Delaying fire growth	Detection of fire
Controlling movement of smoke	Controlling movement of smoke
Means of escape	Controlling fires
Restricting fire spread	Fire fighting
Preventing collapse of structure	

Measures for the control of smoke movement can be partly of a passive nature and partly of an active type. The provision of smoke control doors is a passive measure but the extraction of smoke by mechanical means will come into operation when a fire occurs and a signal is received from the detection system.

The other passive measures are concerned with providing a construction with such surfaces, and where possible with such contents, that a fire will not grow rapidly. The quicker a fire grows less time is available for occupants to escape to a place of safety. Control on linings has been the traditional approach to controlling growth rate. However, it is now (1) recognised that the rate of heat release and the thermal inertia of linings can also be critical factors and some additional control is needed to ensure that the growth rate will be predictably slow. Provision of means of escape requires protected escape routes, smoke control facilities, fire resisting doors and the maintenance of compartmentation. Two purposes of these measures are that the conditions will remain reasonably safe for occupants and sufficient protection is available to move to a place of safety. In large and complex buildings immediate total evacuation will not be possible and the need for places of temporary refuge in the building exists.

The other main purpose of passive measures is to contain a fire and its effects within defined boundaries. This is the main purpose of compartmentation and is achieved by having compartment boundaries which do not collapse in a fire and maintain their integrity. Fire can spread by the collapse of barriers, by the formation of openings which permit hot gases capable of causing ignition passing to the non-fire side and by excessive transfer of heat through the construction such that the materials in contact with the barrier may ignite. Acceptable limits for gaps related to their size and position are specified together with limits on the temperature rise of the unexposed face. There are no statistical data available to show whether fires in practice exploit these mechanisms and whether the limits are correct. Some limited experiments show (2) that the temperature rise limits are very safe.

The whole of a building may be a single compartment if it is small or a building may be divided vertically and/or horizontally into smaller portions. Empirical rules for compartmentations are contained in many regulations and building codes but it is difficult to untangle the precise logic behind these requirements. The sizes of compartments are related to the building occupancy, the building height and the fire resistance of the boundaries. Sometimes the nature of the construction is also taken into account. It is often suggested that a direct relation may exist between the size of the compartment and its fire resistance, i.e. by doubling the fire resistance the size of the compartment may be proportionally increased. An increase in the floor area of a compartment will obviously increase the size of a fire but not necessarily its severity, at least not in the same ratio unless the fire regime changes from fuel bed controlled to ventilation controlled. (Fig. 1)

A more rational basis for the size of a compartment can be the number of occupants at risk or the ability of the fire brigade to fight a fire. Due to the difficulty of attacking a fire in a high rise building it is customary to make every floor a compartment and its fire resistance, ladders. Building codes in North America also allow the sizes to be increased if a building can be approached by the brigade with their appliances on more than one side. An island site can be theoretically four times the size of a building approachable from only one side. May be the compartment dimensions could also be linked with the throw of a hose jet. (Fig. 2)

The association of the fire resistance requirements for different occupancies with their expected fire load seems to be based on the work done by Ingberg (3) in 1928 (Fig. 3). This has often been used to justify the regulatory requirements and sometimes the steps used. Experimental work in the late 50's and early 60's showed that fire severity depended on a number of other factors, most significantly the ventilation conditions (4). The most commonly accepted relationship used for this purpose is,

$$t_f = q_f \times A_f \times k \text{ (min), where}$$

t_f is the fire resistance

q_f is the fire load density (MJ/m^2),

A_f is the floor area (m^2), and

k is a constant.

This relationship is a simplifcation of the actual fire conditions into an equivalent period under the standardised furnace heating conditions which will cause similar damage for specified type of structures.

ACTIVE FIRE PROTECTION MEASURES

Detection of a fire by the use of automatic sensors is an active measure as it functions only when a fire has started. Detectors indicate that a fire has been noticed but will do nothing to control it unless some other system is also activated as a consequence. By noticing a fire at an early stage more time is available for occupants to escape and earlier action can be taken to control the fire. A detection system needs to be coupled with an alarm system, the alarm can be to the management, occupants of the building and to the fire brigade. In the last case there is direct positive action towards controlling a fire. Fire detection signals can be used to function other systems such as closing doors, shutters, escalators, operating smoke extracts or pressurization systems.

The active control of a fire requires the provision of an extinction

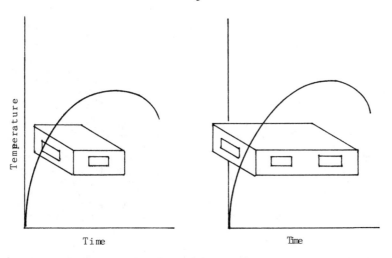

FIG. 1   EFFECT   OF   COMPARTMENT   SIZE   ON   FIRE   SEVERITY

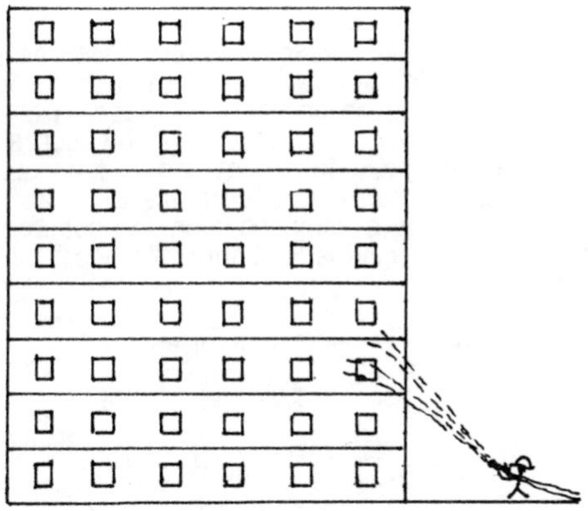

FIG. 2   COMPARTMENTATION   AND   FIRE   FIGHTING

system, the most common example of which is an automatic sprinkler installation. The sprinkler head acts as a detector, opens the orifice through which a spray of water can be applied to the fire underneath. When properly designed and installed the fire will either go out or its rate of growth severely curtailed. Theoretically extinction is the objective but practically the control of the fire size is the main achievement. Statistical data from industrial fires show (5) almost all fires where sprinklers operate are contained within the compartment or the floor of origin.

The sensitivity of the normal glass type or the fusible link type sprinkler heads, with an operating temperature of around 80°C is less than that of a smoke or even a heat detector. The fire is not detected early enough to make a substantial difference to the escape facilities. Recent developments in the design of sprinkler heads have led to the emergence of fast response sprinklers. Their response time is of the same order as heat detectors, hence they can contribute towards life safety as well as fire control (Fig. 4). Sprinkler systems are frequently connected through alarm valves to the fire brigade, this can lead to an early extinguishment of the fire.

Under many building codes, regulations or other legislation in some buildings facilities have to be provided for the access of the fire brigade to the site, and the availability of water supply. In the case of high rise buildings or special risk multi-storey buildings some stairways and lifts may be specially protected to remain available to the brigade. Factors which affect fire brigade contribution are the time at which it is informed, its nearness to the building and ease of access to the seat of fire.

INTER-ACTION

Not only is there inter-action between measures for life safety and property protection but also between some in each category. If a measure influences the fire phenomenon or the time base in any way it will have some consequence for other fire dependent measures. There is at no universal recognition of these inter-actions and no quantification of the influences. A number of countries have made allowances for some of the measures, usually by giving a compensation in passive requirements if an active measure has been provided. Some of these influences are shown below:

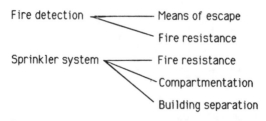

Fire detection by ensuring the awareness of a fire to a predictable early point increases the time available to occupants to escape and therefore they can either travel longer distances or at a slower speed. This can influence the design of escape routes and exit facilities. An early awareness of a fire can also enable an early attack on the fire provided

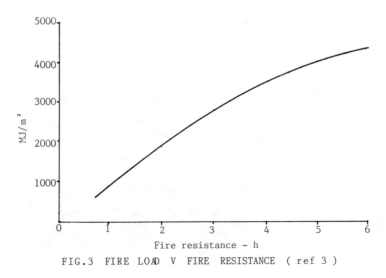

FIG.3 FIRE LOAD V FIRE RESISTANCE (ref 3)

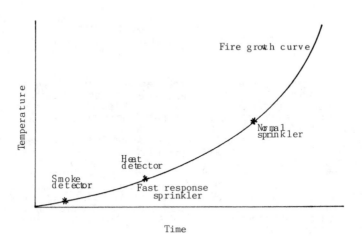

FIG. 4 RESPONSE TIMES OF VARIOUS APPLIANCES

trained personnel are available or the alarm is linked to the fire brigade.

Automatic sprinkler installations have been shown to have a marked effect on the pattern of fires as indicated by an assessment of fire damage or fire spread. (5) The influence on the growth stage can be assumed only if rapid response sprinklers are fitted. Once the system operates it may quickly extinguish a small fire or if the fire seat is concealed it may still control its rate of growth and keep its severity low. With a properly designed, installed and maintained system a significant influence on fire resistance can be foreseen, and as a further consequence on the compartmentation requirements and on the separation specifications to prevent the spread of fire from building to building. It has been suggested that with a sprinkler system a compartment may be of unlimited area except for multi-storey buildings.

The current concessions do not indicate that the authorities have developed any precise rules for this purpose. As the regulatory requirements are expressed in a system of steps, the only possibility is to move one or more steps in the direction of reduced requirements. As many regulatory requirements are applied independent of each other, different steps may be applied in different cases. Only two dimensional interaction is considered at present, if at all.

FIRE SAFETY ENGINEERING

The use of an engineering approach to fire safety is not new, some simple procedures existed 50 years ago which could be considered to be the application of engineering concepts. Fire safety engineering globally applied means an evaluation of risks in a given building and an assessment of the safety measures to see if a net hazard is present. The difference between the two can be construed to indicate the degree of hazard and the design intention would be to reduce the net hazard to zero or a negative value.

Fire safety engineering concepts can be used for sub-systems of the total system. Most progress has been made in the area concerned with fire resistance. It started with a need to extrapolate and interpolate data from standard fire resistance tests and to predict fire resistance which would be obtained in a furnace test. The next development was the quantification of fully developed fires in order to better define fire resistance needs for different buildings. The approach developed after experimental studies (6) takes account of the relevant factors of fire load, ventilation and the compartment boundaries and quantifies the fire severity as an equivalent time in the standard furnace test which would cause the same degree of distress to the structure. A real fire is translated into the furnace test parameters. The impetus for this work came from the steel industry in its competition with concrete to find cost-effective solutions to protecting steel constructions.

These studies highlighted a number of problems which have as yet not been fully resolved. The furnace tests, owing to their simplistic nature, is not capable of reproducing a real fire condition. Some of the main differences between the two are due to the random characterstics of real fires, e.g.

1. The heating conditions in rooms are not uniform,
2. The heat transfer characteristics are different,
3. The environment temperatures fluctuate,
4. The elements are not heated uniformly,
5. Moisture in materials has a delaying effect,
6. Material properties change at high temperature,
7. Building elements interact with each other,
8. Fire duration is usually less than assumed,
9. Fire brigade applies water to heated surfaces,
10. Fire grows at different rates in different parts.

Some of these factors have received attention in the past but others are not so easy to quantify although some account can be taken by introducing adjustment factors. The procedure for estimating fire severity referred to previously has been accepted on an international basis by the CIB(7). It is termed as Method 2 to distinguish it from the standard furnace test, known as Method 1. The difference between these methods is not fundamental but only lie in the definition of the end point, in one case it is given in the regulations, in the other it is obtained from a knowledge of the fire load, etc. In the long term Method 3 should be of greater interest because it attempts to reproduce the heating regime likely to be experienced in practice. This is largely an unchartered territory although a range of idealised relationships has been proposed.

More recently the effect of rate of heating, particularly if it increases rapidly as in fires involving hydrocarbon fuels, on structures has become of some concern. A new relationship known as the "hydrocarbon" curve has been gaining ground in the petro-chemical industry and the associated safety authorities for a more realistic expression of the hazard. Taking $1100^\circ C$ as the maximum temperature of such fires, the time to reach has been shortened from about 90 minutes to 9 minutes. The relationship can be expressed exponentially as below (8) (Fig. 5):

$$T = 1100 [1 - 0.325 \exp (-0.16667 \times t)], \text{ where}$$

T is the temperature in $^\circ C$ and
t is the time in min.

The prediction of fire behaviour of building elements has required the development of computational techniques to determine heat transfer from the heated surface to the inside of the heated element. The transient heating conditions and multi-dimensional heat flow requires use of relationships such as Fourrier equations. For hand calculations these were simplified with assumptions about heat transfer characteristics, surface temperature and the effect of moisture etc. However with the advent of mini and micro computers more elaborate calculations can be made by designers. The best example of the simplified procedure is that given in the code prepared by the ECCS technical committee for the protection of steel structures (9). Of computer programes FIRES-T (10) is well known for calculating heat flow using a finite element technique, whereas Fires-RC calculates the behaviour of concrete elements. Recent developments have produced TASEF program (11) which combines heat transfer and structural calculations and FASBUS (12) which calculates the performance of steel structures.

One of the inputs into such programs is on the properties of materials

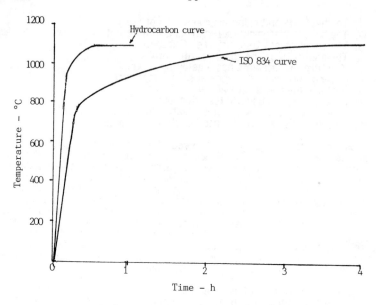

FIG. 5 CELLULOSIC & HYDROCARBON CURVES

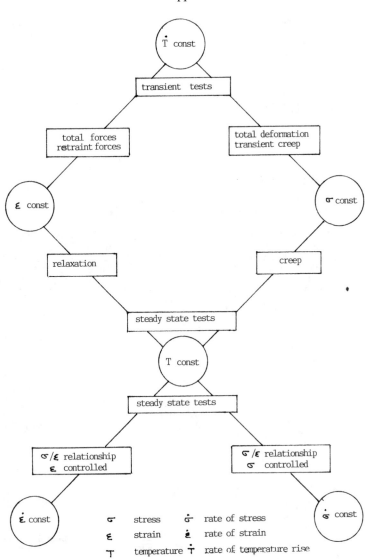

FIG. 6   RILEM MATERIAL TEST PROTOCOL

at high temperatures. Data on such properties in the past had been obtained by testing at fixed temperatures. Analysis has shown that this approach is not realistic as in a fire the temperature variation is progressive and therefore transient heating conditions are more relevant (Fig. 6). This led a RILEM technical committee (13) to produce data on concrete and steel distinguishing between steady state and transient tests and to provide a basis for modelling such properties. The models can be introduced into the main analytical procedure as sub-systems and allow a continuous adjustment of properties at high temperature.

A RATIONAL SYSTEM

Undoubtedly further developments are to be expected in this field. These will have the objective of

a) simulation of a realistic fire environment,
b) simulation of realistic structural behaviour.

It is however necessary to be clear how far can it be practically justified to achieve an absolute degree of realism. We need to bear in mind the needs of legislators, controlling bodies, designers, contractors and material producers. For practical application and for ease of control and verification the system should not be so complex that it becomes the perogative of a few specialists. It must lend itself to application and use by the design profession otherwise its use will not gain favour. Experience with complex calculatin techniques for structural behaviour shows that they tend to be ignored in favour of simpler though less precise methods. One of the objectives of research workers in this field should be to produce a system which is easy to use and gives confidence in its appropriateness to specifiers and approvers.

The wider use of computational techniques requires the regulating authorities to move away from the fixed step approach to safety and permit a more graduated system. The progressive system should also allow the matching of the safety provisions against the specified risks and permit various inter-actions to be taken into account. It is only in this way that a rational approach can be developed.

REFERENCES

1. Thomas, P.H., Bullen, M.L., The importance of insulation in fire growth, Information Paper IP 19/79, Building Research Establishment, Garston, Herts, England, 1979.

2. Schwartz, K.J. & Lie, T.T., Investigating the unexposed surface temperature criteria of standard ASTM E119, Fire Technology, Vol. 21, No. 3, August 1985, p. 169.

3. Ingberg, S.H., Tests of the severity of building fires, US National Fire Protection Association Quarterly, 22, 1928.

4. Law, M., Prediction of fire resistance, Symposium No. 5, Fire Research Station, HMSO, London, 1971.

5. Rogers, F.E., Fire losses and the effect of sprinklers protection of buildings in a variety of industries and trades, Current Paper 9/77, Building Research Establishment, Fire Research Station, Borehamwood, Herts, England, 1977.

6. Pettersson, O., et al, Fire engineering design of steel structures, Publication 50, Swedish Institute of Steel Construction, Stockholm, 1976.

7. Design Guide, Structural Fire Safety, CIB/W14, a proposed publication by the CIB, Rotterdam, 1986.

8. Fire Resistance Tests, Elements of Building Construction, ISO 834-1975, Revision by ISO/TC92/SC2, 1985.

9. European Recommendations for the Fire Safety of Steel Structures, Part 1, Calculation of the fire resistance of loadbearing elements and structural assemblies exposed to the standard fire, Elsevier Scientific Publishing Co. Amsterdam, 1983.

10. Becker, J.M. et al, FIRES-T, A computer program for the fire response of structures, Report No UCB FRG 74-1, Fire Research Group, Department of Civil Engineering, University of California, Berkeley, 1974.

11. Wickstrom, U., TASEF-E, A computer program for Temperature Analysis of Structures Exposed to Fire, Lund Institute of Technology, Sweden, 1979.

12. FASBUS II (Fire Analysis of Steel Building Systems), A computer program developed for the American Iron and Steel Institute, Washington, 1982.

13. RILEM Technical Committee 44-PHT, Properties of materials at high temperatures, Paris, 1985.

THE BEHAVIOUR OF A FIRE

LAW, MARGARET

Ove Arup Partnership
13 Fitzroy Street
London, W1P 6BQ, U.K.

ABSTRACT

The standard fire resistance test and the hydrocarbon fire standard are described. Compartment fires, effective fire resistance and external fire exposure are reviewed. The practical experience of building fires and the contribution of active fire protection measures are summarised.

INTRODUCTION

Building fires affect the way structures behave because they change the physical and mechanical properties of the materials of construction. In assessing the structural fire protection needed, an estimate of the potential fire exposure - the rate of heating, the fire temperature, the fire duration - ought to be taken into account. In practice, the standard of protection is usually prescribed by building regulations or insurance company rules and these define performance in terms of survival time in the standard fire resistance test. This paper describes standard fires, experimental fires and fires in practice.

STANDARD FIRE

In a standard fire resistance test an element of structure, loaded or not as appropriate, is placed in a furnace with the fuel input controlled so that the temperature of thermocouples adjacent to the heated surface follows a standard temperature-time curve given by the following equation:

$$T - To = 345 \log_{10}(8t + 1) \qquad (1)$$

where t = time   (min)

        T = furnace temperature at time t (°C)

      To = initial furnace temperature   (°C)

The heat transfer coefficient for convection can be taken as 25 W/m^2 °C . For heat transfer by radiation, the resultant emissivity of the flames, combustion gases and exposed surfaces in the furnace can be taken as 0.5. (1) (2). For well insulated steel elements, the heated surface temperature, T_s, can be taken as equal to T (1). For concrete elements T_s can be taken as about 0.85T.(2). In general, a conservative solution will be obtained in heat transfer calculations if it is assumed that non-metallic surfaces exposed to the fire are at the temperature T given by equation (1).

Although the temperatures attained in the standard test are reasonably representative of those attained in most building fires, that is of order 1000°C, the test does not represent the fire behaviour likely to be experienced in practice. It can identify the likely mode of failure of a heated element, but its main use has been to rank or grade elements in order of performance for code purposes. The relationship between the standard fire resistance test and building fires is discussed later.

The standard test does not reproduce the high heating rates which may develop in some fires in industry or on oil platforms, when significant amounts of liquid or gaseous fuels may be burning. The so-called Mobil curve has been developed (3) which attains 1000°C within 10 minutes and the British Fire Research Station has recommended a similar standard (4).

COMPARTMENT FIRES

Experiments have been carried out, on a variety of scales, designed to measure various aspects of fire behaviour in single compartments. Usually all the fuel was ignited to burn simultaneously and was allowed to burn out completely. In early developments, the temperature-time curves obtained were compared with equation (1) and a relationship between fire load per unit floor area $\frac{L}{A_F}$ ($kglm^2$) in wood equivalent and the furnace time, t, was derived. Later work explored the importance of ventilation, two kinds of fire behaviour being identified : free burning and ventilation controlled (5) (6). The rate of burning of the fuel R (kg/s) for the ventilation controlled condition was given by :

$$R \simeq 0.1 \ A\sqrt{h} \qquad\qquad (2)$$

where A = area of ventilation opening (m^2)

h = height of ventilation opening   (m)

The ventilation opening not only admits air into the compartment but also offers a path for heat release. The temperature and duration of the fire is affected by the ventilation, feedback to the fuel, the amount of fuel, and the heat absorbed by surrounding surfaces. A balanced condition is usually attained which continues until the fuel becomes exhausted.

Later experiments showed that $R/A\sqrt{h}$ is not constant; it depends on the parameter $(A_t-A)/A\sqrt{h}$ $(m^{-\frac{1}{2}})$, where A_t is the area of the internal envelope, and on the depth to width ratio of the compartment. The temperature attained depends on the value of $(At-A)/A\sqrt{h}$ (7), being a maximum when it is about 15.$(m^{-\frac{1}{2}})$.

The rate of burning of well ventilated fires depends on the amount and type of fuel. The effective fire duration depends on the type, being about 20 minutes for conventional furniture.

Calculated temperature-time curves of a compartment fire can be derived by making a heat balance. The most comprehensive information published (8) assumes a rate of burning given by equation (2) and gives a series of curves for different values of the 'opening factor' $A\sqrt{h}/A_t$, for different fire loads, (converted to wood equivalant) and with compartment walls of different thermal properties. It may overestimate the temperature and underestimate the duration of the low fire load, well ventilated fires; this effect is only likely to have significance when the structure is unprotected steelwork, the results tending to be conservative. The convective heat transfer coefficient is taken as $20W/m^2°C$ and the resultant emissivity is taken as 0.7 for columns and deep beams and 0.5 for shallow beams (depth less than half the width). This work has been criticised for assuming a uniform temperature distribution, whch would only be expected in small compartments, and for assuming a mainly wood-type fire load. Nevertheless, it is a significant advance and gives representative structural temperatures for many types of building fire.

EFFECTIVE FIRE RESISTANCE

In order to illustrate the importance of the ventilation, fire-load and heat balance of the compartment fire, a relationship between these and the standard fire resistance test has been devised (9). It gives the exposure time te (min) in the standard test which would give the same effect, in terms of a critical temperature of a protected steel element or a reinforced concrete element, as would the compartment fire:

$$te = K \frac{L}{\sqrt{A(At-A)}} \qquad (3)$$

where K = constant $\simeq$ unity

L = fire load in wood equivalent   (kg)

EXTERNAL FIRE EXPOSURE

Flames emerging from windows and other ventilation openings can heat external structural elements but the exposure is usually less severe because the elements are free to cool to the ambient air, Following fundamental work by Yokoi (10) an analysis of experimental data (11) gave

the following equation for the flame height, z(m) above the top of a
window of width w (m):

$$z + h = 12.8 \left(\frac{R}{w}\right)^{2/3} \qquad (4)$$

Using the heat balance approach and the more differentiated estimate
of rate of burning, a model of flame projection and heat transfer to
external elements has been developed (11) which is used for design
manuals (12) (13).

PRACTICAL EXPERIENCE OF BUILDING FIRES

Most building fires are small. Even when they grow too large for
the fire brigade to extinguish, the cooling of the structure by hose
streams has a beneficial effect. Statistical data for all fires attended
by fire brigades show the proportion of fires which exceed a given size
and the beneficial effect of automatic detection and automatic extinction
measures in reducing the chances of a fire becoming large. Analysis of
these data also shows how fires tend to grow exponentially, (14, 15, 16,
17). On average, automatic detectors can result in a 50% reduction in
damage. Automatic sprinklers can reduce the chance of a large industrial
fire by a factor 6 and a large shop fire by a factor 2.

Burning rates and fire durations have been assessed for a variety of
industrial fires, with and without sprinklers, attended by fire brigades
(18). The annual probability of a fire occurring has also been assessed
for a range of occupancies (19).

When there are a number of alternative protection measures available
in a building it would be reasonable to assume that it is not always
necessary to provide fire protection to survive a total burnout.

In certain buildings, notably well ventilated car parks, the
potential fire exposure is so low that structural protection is not
usually needed, provided that the construction itself is
non-combustable.

CONCLUSIONS

In order to decide whether a structure needs protection and if so
how much, the potential fire exposure should be assessed taking into
account the type of fuel (wood or hydrocarbon for example), the amount of
fuel and how much is likely to be burnt, the heat balance (ventilation,
heat losses and feed back) and the effectiveness of various fire
protection measures.

REFERENCES

1. European Convention for Constructional Steelwork.
    European Recommendations for the Fire Safety of Steel Structures
    Elsevier, 1983.

2. Comite Euro-International du Beton Design of Concrete Structures
    For Fire Resistance Bulletin d'information No 145, Paris, 1982.

3. Warren, J. H. and Corona, A.A. This Method Test Fire Protective
    Coatings. Hydrocarbon Processing January, 1975

4. Shipp, M. Hydrocarbon Fire Standard for Offshore Installations.
    Fire Surveyor December 1984.

5. Fujita, K. Characteristics of fire inside a non-combustible room
    and prevention of fire damage, Japanese Ministry of Construction
    Building Research Institute Report 2(h) Tokyo

6. Thomas, P.H., Heselden, A.J.M. and Law, Margaret. Fully developed
    compartment fires - two kinds of behaviour. Fire Research Technical
    Paper No 18, HMSO, London 1967.

7. Thomas P.H. and Heselden A.J.M. Fully developed fires in single
    compartments (CIB Report no 20) Joint Fire Research Organization
    Fire Research Note 923 Borehamwood 1972.

8. Pelterson, O., Magnusson, S-E,and Thor J. Fire engineering design
    of steel structures, Publication 50, Swedish Institute of Steel
    Construction, Stockholm, 1976.

9. Law, Margaret. Prediction of fire resistance Paper No.2. of
    Symposium no.5. Fire resistance requirements for buildings-
    a new approach, London, HMSO, 1973

10. Yokoi, S. Study on the prevention of fire spread caused by hot
    upward current. Report of the Building Research Institute No.34
    Tokyo 1960.

11. Ove Arup and Partners. Design guide for fire safety of bare
    exterior structural steel. Technical Reports. American
    Iron & Steel Institute, Constrado, London 1977.

12. Fire safe structural steel, A design guide Washington D.C.,
    American Iron and Steel Institute, 1979.

13. Law, Margaret and O'Brien, Turlogh: Fire safety of bare external
    structural steel, Croydon, Constrado, 1981.

14. Rogers, F.E. Fire losses and the effect of sprinkler protection
    of buildings in a variety of industries and trades. Building
    Research Establishment Current Paper CP 9/77 February 1979.

16. Ramachandran, G. <u>Building Research Establishment IP 27/80</u>

17. Morgan, H.P. and Chandler, S.E. Fire sites and sprinkler effectiveness in shopping complexes and retail premises. <u>Fire Surveyor</u> October 1981.

18. Thomas P.H. and Theobald, C.R. Studies of fires in industrial buildings - Part 2: The burning rates and duration of fires <u>Fire Prevention Science and Technology 17</u> (1977) 15-16.

19. Rutstein, R. The estimation of the fire hazard in different occupancies <u>Fire Surveyor</u> April 1979.

PROBABILISTIC CONCEPTS IN FIRE ENGINEERING

M. KERSKEN-BRADLEY
consulting engineer, Munich, Germany, F.R.

Abstract

As a prerequisite for the application of probabilistic concepts, the first part of this contribution deals with models
used for the verification of an adequate structural fire resistance referring to the heat exposure and the structural
response. In accordance with the CIB design guide for Structural Fire Safety three methods of assessment are presented.
It is emphasized that these three methods do not reflect com-
peting scientific concepts, but rather represent consecutive
steps of accuracy in modelling.
In its second part probabilistic concepts are dealt with,
again briefly outlining three approaches which, however, dif-
fer considerably with regard to their background. The first is
the Structural Fire Safety Concept, as developed in Germany,
F.R. and suggested in the CIB design guide; the second is the
Swiss Risk Evaluation Method and the third is a system approach
based on logic tree analysis, mainly promoted in the United
States. An attempt is made to compare these three concepts
with regard to some common features.

0. NOTICE

The author would like to mention that the material of this contribution has already been presented at various occasions and
is published e.g. in /1,2,3,4/. This paper is basically an ed-
itorially improved and updated version of publication /5/.
Since e.g. /6/ gives a very detailed description of a probability based structural design concept, the contribution gives
only a rough outline of the decisive features and pursues some
comparative discussions.

1. INTRODUCTION

Probabilistic concepts can only be appreciated if the engineer's task is understood in a sufficiently comprehensive manner. In this sense, structural fire design calls for safe and economic design solutions comprising the following aspects:
- The basic design of a building with regard to structural fire protection including an adequate arrangement of the building by providing accessability, rescue routes, compartmentation and arrangement of fire barriers and also an appropriate choice of building materials.
- Structural detailing including choice of an appropriate structural system, detailing of components, supports, joints.
- Verification of an adequate structural fire resistance using models for assessing the heat exposure of the structure in a compartment fire and models for assessing the structural response.
- Guidance on decisions with regard to the adequacy of structural provisions by assessing the fire risks and the resulting interaction between structural provisions and risk considerations.

This contribution focusses on the two latter aspects, mainly referring to the state of art, as it is relevant for practical application.

2. HEAT EXPOSURE AND STRUCTURAL RESPONSE

2.1 Definitions

The heat exposure model specifies the design situation to be considered in terms of a fire, affecting
- the structure as a whole or
- only a limited part of the structure, e.g. a defined fire compartment

and the corresponding thermal actions represented by temperature-time curves for the entire fire process or a limited part of it.

Structural models specify the structural system to be considered which may be
- the structure as a whole or

- subassemblies or individual members with corresponding support and restraint conditions
as well as the relevant material properties and mechanical
loads.

Verification is performed by checking the resulting structural
response with regard to prescribed performance criteria relating to
- the load bearing capacity (strength, stability, ductility)
- the separating function (thermal insulation, integrity to
  fire penetration).

Within Europe, and actually, world-wide, three approaches may
be distinguished which mainly refer to different heat exposure
models /6,7/. However, it can easily be shown that these three
approaches actually represent consecutive steps of improved
modelling rather than different scientific concepts.

2.2 Standard Fire Exposure
- Assessment Method 1 -

In the traditional approach reference is made to a heat exposure model represented by a standard temperature-time curve,
generally the national variant of ISO 834. This curve describes
a monotonically increasing temperature in time. Structural requirements are expressed by a duration of exposure to this temperature-time curve during which structural elements have to
fulfill specified performance criteria. This defines the required fire resistance.

Within the traditional approach, the required fire resistance
is generally stipulated in building codes or regulations. Differing physical conditions governing the heat exposure of the
structure are not accounted for or only implicitly in conjunction with risk considerations by a fairly rough rating of buildings (cf. also sec. 3.1):

$$\text{requ. } t_f \sim t \text{ (occupancy, building height,}$$
$$\text{building component function, ...)} \tag{1}$$

With regard to the structural model, verification was originally
confined to an experimental evaluation of the fire resistance or

by reference to approved catalogues:

$$t_f \text{ (component test)} \geq \text{requ. } t_f \text{ (component)} \qquad (2)$$

Gradually, an analytical verification is accepted as an alternative to testing the load bearing capacity and for some cases the thermal insulation /8,9/. This allows an assessment of components and subassemblies with arbitrary restraint conditions or even the entire structure, which is well beyond the possibilities of the original rating of individual components by testing.

In analytical evaluation and, subsequently, also in experimental evaluation (e.g. /10/), allowance is made for the actual degree of mechanical loading instead of assuming the conventional service load. These features indicate a considerable improvement of the structural model which, in some cases, may be in excess of the accuracy of the heat exposure model and may possibly be well beyond the intentions of the original rating of components, i.e.

$$t_f \text{(structure)} \quad \text{vs.} \quad \text{requ. } t_f \text{(component)}$$

A first step towards improving the heat exposure model is by introducing code requirements, which explicitly allow for a dependency between the required fire resistance and the fire loads within the fire compartment considered (e.g. The Netherlands):

$$\text{requ. } t_f \sim t(\ldots, \text{ fire load density, } \ldots) \qquad (3)$$

2.3 Equivalent Time of Fire Exposure
— Assessment Method 2 —

Introducing not only the fire loads but also other relevant factors affecting the heat exposure of the structure in the case of fire, as are ventilation conditions, thermal properties of the structure enclosing the compartment and possibly the combustion behaviour, leads to the model of an equivalent time of fire exposure:

$$t_e \sim t \text{(physical conditions)} \qquad (4)$$

e.g. in the form used in /11/ and also suggested in /6,7/

$$t_e = q \cdot w \cdot c \cdot m \text{ (min)} \tag{5}$$

where

    q   fire load density (MJ/m^2)

    w   ventilation factor

    c   conversion factor $(min\ m^2/MJ)$

    m   combustion factor - if considered.

This model relates the effect of the expected (natural) compartment fire to the effect of a standard fire exposure. In its
broadest sense, the equivalent time of fire exposure may be interpreted as a comparative measure for rating the expected fire
severity in a compartment with regard to relevant physical conditions. In this sense, the equivalent time of fire exposure is
independent of the type of construction and material and of the
specific component and performance criteria of concern:

$$t_e \sim \text{(measure of fire severity)} \tag{6}$$

Consideration of appropriate safety factors (cf. sec. 3.2) renders the required fire resistance of the structure and its
structural components, which can easily be compared with traditional requirements (cf. sec. 2.2). Verification can be performed by an analytical verification of the fire resistance or
by reference to test results or established catalogues. Hence,
the information and experience from many years of fire resistance tests can be utilized.

$$\text{requ. } t_f \sim t \text{ (fire severity, safety factors)} \tag{7}$$

This model is used e.g. in Germany, F.R., for an assessment of
industrial buildings /11/. Similar models are used in the CSSR
or are promoted in Canada /12/.

By introducing categories of fire severity for fire compartments, this model may be used not only as an alternative to
current (traditional) code requirements, but as a basis for deriving physically based code requirements.

A scientific improvement of the model may be achieved by introducing a dependency between the equivalent time of fire duration

and the thermal and structural properties of the various structural members considered in the fire compartment. Instead of a general measure for rating the fire severity to be expected in a fire compartment, the equivalent time of fire exposure could then be interpreted as an actual temperature action effect, expressed in the standard time domain

$$t_e \sim \text{(temperature action effect)} \tag{8}$$

2.4 Compartment Fire Exposure
- Assessment Method 3 -

Basically, accounting for the same factors as the equivalent time of fire exposure, but relating them via heat and mass balance equations, renders the temperature-time curve of the expected (natural) compartment fire. From this curve the temperature action effects in the structure are calculated. Verification of the load bearing capacity criterion is performed as for a normal structural design by an analytical evaluation of the resistance capacity (R) as a function of time, and comparison with the relevant mechanical load effects (S) for an accidental design situation - including appropriate safety factors (cf. sec. 3.2)

$$\min R(t) \geq S(t) \tag{9}$$

with a corresponding formulation for insulation criteria, if possible. The model is not applicable if integrity criteria are decisive. For structures which may be sensitive to the rate of heating and for fire loads associated with a high heating rate, the greater accuracy of this model may be essential. For common data sets of physical characteristics, temperature-time curves can be prepared as design aids, thus, considerably reducing the calculation effort.

This model is used e.g. in Sweden /13,14/, but also world-wide for checking simplified heat exposure models.

2.5 Limitations

The heat exposure models currently used for design refer to the fire compartment as a basic unit and assume a full fire involvement of the compartment (flash-over) with a ventilation controlled combustion and a uniform temperature distribution. The validity of these assumptions may be questioned for extremely concentrated fire loads, very large fire compartments and for extreme ventilation conditions including wind and cross-ventilation.

These limitations should not impede the use of advanced models as compared to an exclusively empirical assessment, but they should be considered in the general design and detailing of the structure. Eventually, more refined scientific models are under investigation for deriving simple design guidelines.

3. PROBABILISTIC MODELLING

3.1 Status Quo

The traditional approach accounts for risk considerations by the aforementioned rating of structural members according to their function and of buildings according to their occupancy and height in view of possible fire hazards (cf. rel.(1)). As a generally valid prescription, it is a priori conservative and may lead to uneconomic designs in a variety of applications.

An essential step towards a clearer risk differentiation is accomplished if a reduced fire resistance - e.g. by one class - is accepted if sprinkler systems are employed. This is not an unusual arrangement in many countries, at least as concerns an individual assessment of projects.

$$\text{requ. } t_f \sim t \ (\ldots, \text{ sprinkler system, } \ldots) \tag{10}$$

This step is essential, because it introduces the notion of fire risk not only with regard to possible hazards, but also with regard to the frequency of fires.

3.2 Structural Fire Safety Concept

3.2.1 Object

The aforementioned arrangement, i.e. reducing the required fire resistance by e.g. one class if sprinkler systems are installed, is generally based on an empirical risk assessment. An attempt to quantify the reduction of structural fire risks in view of non-structural protection measures, is by introducing a probability based design concept /6,7,15/. This concept considers the following features

i.   the frequency of fire occurrences, severe enough to cause structural damage (p_a) depending on
     . the occupancy and size of the fire compartment
     . the efficacy and long-term reliability of measures for detecting and fighting fires

ii. safety requirements in terms of tolerable lifetime failure probabilities for structures (p_f), reflecting the safety considerations of the community in view of potential hazards due to structural failure, thus depending on
     . the use of the building, its location and size
     . the size and location of a fire compartment within the building
     . the function of the various structural components
     which - to a great extent - correspond to those features also considered in an empirical assessment, but are tentatively quantified in terms of probabilities.

In addition

iii. the uncertainties in modelling including the randomness of physical and mechanical variables

are explicitly accounted for.

Probabilistic analysis renders tolerable conditional failure probabilities (or failure rates):

$$p_{f,a} = \frac{p_f}{p_a} \tag{11}$$

applying to the conditional event "if a severe fire occurs",

considering the probability for this event. On the basis of
equ. (11), safety factors γ for the verification of an adequate
fire resistance are derived using simple reliability methods,
cf. e.g. /6,7,15/. For an assessment based on the equivalent
time of fire exposure (cf. sec. 2.3), this renders

$$\text{requ. } t_f = t_e \; \gamma \; \gamma_n \tag{12}$$

where

γ accounts for potential structural hazards
and the uncertainty in the assessment

γ_n accounts for a reduced fire frequency due to
special detecting and fighting provisions.

In a corresponding manner safety factors may be allocated to
equ. (9). In the appendix of /7/ the following factors γ_n are
tentatively suggested:

No.	fire detecting and fighting provisions	γ_n *)
1	average standard public fire brigade	0.8
2	adequately maintained sprinkler system	0.6
3	acknowledged private fire brigade	0.7...0.9
4	adequately maintained detection and alarm systems	0.8...1.0

*) when considering all provisions, a lower limit for γ_n should
be observed, tentatively specified as $\gamma_n \geq 0.4$

Table 2.1   Factors γ_n for assessing fire detecting and fighting provisions

3.2.2 Discussion

Whilst the dependency of structural fire requirements on potential structural hazards is straightforward, the dependency on
non-structural measures (governing the frequency of severe
fires) is not yet generally acknolwedged as design parameter.
The major argument brought forth refers to the reliability of
these measures in the sense that, if e.g. a sprinkler fails to
suppress an initial fire, then a reduced fire resistance of the
structure will exhibit a considerable hazard, especially for

the fire brigade. This argument is correct, provided the condition "if" is pursued consistently:

Imagine 100.000 building fires p.a. within a certain geographical region; assuming that no building is equipped with sprinklers, maybe 10.000 fires will develop to fires, severe enough
to damage the load bearing or partitioning structure - mainly
due to the time elapsed before manual fire fighting commences.
That means that in 10.000 fire incidents structural failure is
a potential hazard. Depending on the required fire resistance
of the structure stipulated in codes, structural failure will
occur in N_1 out of 10.000 cases, implying an accepted failure
rate (failure probability) of $N_1/10.000$ and imposing N_1 risk
situations on fire fighters.

And now imagine the same geographical region with 100.000
building fires p.a., but imagine that all buildings are
equipped with sprinkler systems. Assuming a certain degree of
reliability for these systems, maybe 2000 fires are not sup-
pressed by the sprinkler system alone and may develop to fires,
severe enough to damage to load bearing and partitioning structure - if not controlled by the fire brigade. Not considering
the remaining possibility of a timely manual suppression of
fires, this means that, in 2000 fire incidents, structural failure is a potential hazard. Depending on the fire resistance accepted for these buildings, structural failure will occur in N_2
out of 2000 cases, implying a failure rate (failure probability)
of $N_2/2000$ and imposing N_2 risk situations on fire fighters.

If N_1 risk situations are accepted by fire fighters for buildings without sprinklers, the same number should be acceptable
for buildings with sprinklers: $N = N_1 = N_2$. Hence, the acceptable failure rate of the structure is approximately five times
higher for buildings with sprinklers as compared to buildings
without:

$N/10.000$ versus $N/2000$

This shows clearly that a considerably lower fire resistance
should be acceptable, if sprinklers are installed. It should be
noted that in this game of numbers, no successful fire brigade

action in the wake of sprinkler failure is assumed, implying no operations within the building. If limited inside operations are considered, in far less than 2000 fire incidents structural failure will be a potential hazard.

Another argument brought forth regarding the consideration of protection measures governing fire frequencies, is the poor data base for evaluating these frequencies. This does not hold for sprinklers, but may apply to other measures which, at present, can only be judged in relation to sprinklers. There are also some difficulties associated with the assessment of the frequency of fire occurrences as a function of e.g. the size and occupancy of the compartment. But, on the other hand, calibration to design results which are broadly agreed, will help avoiding excessively wrong estimates. However, it should be noted that investigations are being carried out e.g. presently in Germany, so that improved information can be gradually adopted.

3.3 Swiss Fire Risk Evaluation Method

3.3.1 Object

The Structural Fire Safety Concept described in section 3.2 is limited to risks associated with structural failure. A much broader approach is taken in the Swiss Evaluation Method /16/, derived from an insurance rating procedure.

Herein a normalized measure of fire risk (R) is calculated which may - by interpretation by the author - be traced back to an assessment of the increase or decrease of the expected losses (E_L) versus an average loss $(\overline{E}_L)$ depending on the specific features (X_i) of an individual project. This interpretation renders

$$E_L \cong \overline{E}_L + \Sigma (X_i - \overline{X}_i) \frac{\partial E_L}{\partial X_i} \tag{13a}$$

$$R \cong \pi \exp(X_i - \overline{X}_i) \frac{\partial E_L}{\partial X_i} = \pi x_i = \frac{P}{NSF} A \qquad (13b)$$

where the factors x_i describe the relative influence of the feature X_i on the measure of risk. P combines those factors relating to the potential fire risk as are
- the amount and type of fire loads including their combustion behaviour, toxiticity and smoke development
- the influence of height and size of the building and spatiousness of the compartment.

N represents all factors relating to the possible inadequacy of standard fire protection measures as are manual extinguishers, water supply and hydrants, etc. S combines those factors relating to special fire protection measures with regard to detection, alarm and force of fire brigades as well as automatic extinguishing systems. F comprises factors f_i relating to the fire resistance of the structure and the size of sub-compartments. A , finally denotes the relative influence of the occupancy in terms of fire frequencies.

This measure of risk (R) is then compared with a tolerable risk, specified as R = 1.3 and which is pondered by the number of people endangered in the case of fire $(p_{H,E})$

$$R \leq R_u = 1.3 \ p_{H,E} \qquad (14)$$

3.3.2 Discussion

Due to the very different background of this method, a comparison with the Structural Fire Safety Concept is not straightforward.

An attempt for a comparison is made in the following, which may be in excess of the intentions of those who developed the Risk Evaluation Method, but nevertheless gives some interesting insights.

A factor f_1 in /16/ denotes the influence of the fire resistance of the load bearing structure and is specified as

$$f_1 = \left\{ \begin{array}{lll} 1.0 & \text{for} & t_f < F30 \\ 1.2 & \text{for} & t_f = F30/F60 \\ 1.3 & \text{for} & t_f \geq F90 \end{array} \right.$$

and may be written as

$$f_1 = 1 + t_f/300 \leq 1.3 \tag{15}$$

It is interesting to note that fire resistances exceeding F90 are not accounted for as a further risk reducing feature. Equ. (15) allows a tentative comparison with the Structural Fire Safety Concept applied to assessment method 2 (cf. equ. (12)):

$$\text{requ. } t_f = 300 \left(\frac{P}{R_u \ N \ S \ f_2 \ \cdots} - 1.0 \right) \leq 90 \tag{16}$$

rendering a γ_n-value for special fire protection measures as

$$\gamma_n = \frac{P/S - R_u \ N \ f_2 \ \cdots}{P - R_u \ N \ f_2 \ \cdots} \tag{17}$$

Sprinkler systems alone are rated as $S = 1.2 \cdot (1.75 \ldots 2.0)$ wherein the value 1.2 rates the detection effect, resulting in $\gamma_n = 0.42$ as an _upper_ bound in /16/ (disregarding the limit of F90) as compared with $\gamma_n = 0.6$ in section 3.2. In this context, however, it should be mentioned that application of this method in Switzerland is presently intended only in conjunction with minimum requirements for the fire resistance which, however, are lower than traditionally required.

Another interesting aspect is the rating of the fire load density. P in equ. (13b) or (16) comprises the rating factor $\hat{q}$ taking values from 0.6 to 2.5 depending on the fire load density q (in MJ/m²), which may be related via

$$\hat{q} = 1 + 0.29 \ln\left(\frac{q}{200}\right) \leq 2.5 \tag{18}$$

Hence, requ. t_f is a function of $\ln(q)$ instead of q thus suggesting a weaker rating than in assessment method 2 or 3. With regard to questions of validity and data base, the Swiss Fire Risk Evaluation Method has the advantage of not claiming to be a scientifically based method, but only a commonly agreed empirical ponderation procedure. Hence, the question of validity is superfluous. However, by reference to equ. (13), a scientific basis for this method may be developed in the future.

3.4 Logic Tree Analysis

3.4.1 Object

Within the fire risk assessment, logic tree analysis is an important tool which may be used on various levels of sophistication in modelling.

Logic tree analysis is concerned with an investigation of the possible extent of fire development within the room of origin and the spread of smoke and flames beyond successive fire barriers, considering the various features governing these processes.

The most common presentation is by reference to the sequence of events leading to a certain hazard, which may be
- a physical phenomena, e.g. full fire involvement of a building /12/ or a fire reaching critical areas of a building or plant /18/, considering various initial fire events
- possible consequences from physical phenomena, e.g. a multiple fatality fire /19/

rendering so-called event trees or fault trees. Alternatively, the sequence of strategies for preventing a certain hazard may be pursued, rendering so-called decision trees /20/.

An analysis may be

i.   qualitative (non-numerical) as an aid for ensuring that all hazards are identified and no essential event sequence is omitted when specifying the fire protection system of a building

ii.   supplemented by an empirical judgement for the likelihood of event sequences including the efficacy of protection measures

iii.   or supplemented by an assessment of probabilities for

event sequences including the reliability of protection
measures, which may be performed
. on the basis of subjective judgement or
. on the basis of statistical data, physical models
  and probabilistic analysis.

With regard to physical phenomena related to fire development
and spread, an assessment of probabilities for event sequences
can be presented by probability curves /17,21/, based on H.E.
Nelson's Fire Safety Design Methology. Such curves, also denoted as L-curves, may describe, e.g.

- the probability for a certain portion of the room of fire
  origin to be seized by fire, considering the possibility of
  self-termination, manual suppression, automatic suppression.
  This renders curves which decrease with increasing portion of
  the room considered, approaching the probability for a full
  fire involvement of the room
- the probability for a barrier to sustain exposure to fire
  with increasing time of exposure
- the probability for successive barriers to be surpassed by
  smoke or flames, decreasing with the number of barriers and
  approaching the probability for a full fire involvement of
  the building.

It is then suggested /17,21/ to compare these curves with tolerable levels reflecting the safety considerations of the community.

3.3.2 Discussion

In a qualitative presentation, logic tree analysis corresponds
to the common sense approach for dealing with complex problems.
Applied in a formalized manner, it is an essential tool for
avoiding weak points in the fire protection system. It is highly recommended to develop simple methods of assessment and
presentation, similar to those procedures under development in
the United States, but considering the European fire protection
tradition. On a pre-codification level, corresponding attempts
are in progress with the objective to result in a format as

simple as in the Swiss Risk Evaluation Method /16/.

With regard to structural fire safety, fault tree analysis e.g. for multiple fatality disasters clearly identifies the contribution of the fire resistance of the structure to the possibility of avoiding this hazard: The contribution of often negligible (cf. also /16/, Appendix 3).

It should be noted that the Structural Fire Safety Concept also uses the idea of simplified event sequences /6,22/ for determining the probability of a full fire involvement of the fire compartment (cf. p_a in sec.3.2). Hereby, the probability for the events: "fire occurrence, failure of manual and automatic suppression" are considered, rendering the tolerable conditional probability for the event "structural failure".

The very extensive use of logic trees, corresponding networks and evaluation charts suggested in /17/ mainly refers to a subjective judgement of probabilities at present (cf. also /23/). However, gradually improved statistical data, physical models and probabilistic models may be introduced. It, nevertheless, may be interesting to note that the probability for a successful automatic suppression is specified as 0.8...0.98 and suppression by the public fire brigade is rated by probabilities 0.1...0.4 in /17/, the corresponding probabilities in /7/, providing the basis for table 2.1 are 0.98 for sprinkler systems and 0.9 for the fire brigade respectively, which is either too optimistic or reflects a considerably higher standard of European fire brigades.

4. CONCLUDING REMARKS

Design models as well as probabilistic concepts applied for the assessment of an individual project are often questioned with regard to the problems associated with a change of occupancy. This applies both, to an alteration of physical conditions and risk considerations. This aspect definitely deserves attention and may be considered by the following provisions:

- the main structural components constituting fire compartments (fire walls) could be designed irrespective of favourable

physical conditions and special fire protection measures due to the anticipated occupancy
- if an alteration of occupancy is rather probable, the input-data should be chosen in a conservative manner
- in case of a significant change of occupancy or in case of structural alterations, reassessment is required; this aspect should be brought to the attention of the client.

This drawback, however, should not impede the application and development of methods for ensuring optimum fire protection systems for buildings, optimal with regard to safety and economy.

5. REFERENCES

/1/   BUB, H., KERSKEN-BRADLEY, M., SCHNEIDER, U.; Structural Fire Protection Levels for Industrial Buildings; American Concrete Institute (ACI), Fall Convention, San Juan, 1980

/2/   BUB, H., HOSSER, D., KERSKEN-BRADLEY, M., SCHNEIDER, U.; Eine Auslegungssystematik für den baulichen Brandschutz, BRABA, Erich Schmid Verlag, H. 4, 1983

/3/   KERSKEN-BRADLEY, M.; A Safety Concept for Structural Fire Design, 6. IBS Seminar Karlsruhe, 1982, Vol. I, pp. 155-170

/4/   KERSKEN-BRADLEY, M.; A Probabilistic Safety Concept, CCE/CEC/KEB, International Conference on Fire Safe Steel Constructions. Luxembourg 1984, Acier 49, 3, 1984, pp. 123-126

/5/   KERSKEN-BRADLEY, M.; Safety Concepts/European Fire Engineering, European Seminar on Fires in Buildings, Luxembourg 1984, Elsevier, pp. 213-222

/6/   CIB W14 Workshop Report 'Structural Fire Safety', January 1983; Fire Safety Journal Vol. 6 (1983)

/7/   CIB W14 Design Guide 'Structural Fire Safety', to be published 1986

/8/ ECCS, European Recommendations for the Fire Safety of Steel Structures Technical Committee 3: Fire Safety of Steel Structures, July (1981)

/9/ CEB, Design of Concrete Structures for Fire Resistance, Appendix to the CEB/FIP Model Code, Bulletin d'Information No. 145, Paris (1982)

/10/ DIN 4102 Brandverhalten von Baustoffen und Bauteilen, Beuth Verlag, Berlin (1981)

/11/ DIN 18230, Brandschutz im Industriebau, Vornorm, 1982, Beuth Verlag, Berlin

/12/ MEHAFFEY, J.R., HARMATHY, T.Z.; Assessment of Fire Resistance Requirements, Fire Technol., 17 (4) (1980)

/13/ National Swedish Board of Physical Planning and Building; Fire Engineering, Design; Comments on SBN, No. 1, 1976

/14/ MAGNUSSON, S.E., PETTERSSON, O.; Rational Design Methology for Fire Exposed Load Bearing Structures, Fire Safety J., 3 (1981)

/15/ DIN, NABau, Grundlagen zur Festlegung von Sicherheitsanforderungen im baulichen Brandschutz, Beuth Verlag, Berlin (1979)

/16/ SIA Recommendation: Brandrisikobewertung, issued by SIA, VKF and BVD, Draft April 1983

/17/ FITZGERALD, R.W.; Building Fire Safety Evaluation, Worcester Polytechnic Institute, Center for Fire Safety Studies, Workbook 1984

/18/ APOSTOLAKIS, G.; Some Probabilistic Aspects of Fire Risk Analysis for Nuclear Power Plants; First International Symposium on Fire Safety Science, Gaithersburg, ML, 1985

/19/ RASBASH, D.J.; Analytical Approach to Fire Safety, Fire Surveyor 9 (4) (1980) 20-33

/20/ NFPA Decision Tree, Committee on Systems Concepts, 1974

/21/   Interim Guide for Goal Oriented Systems Approach to
        Building Fire Safety, Appendix D, Building Fire Safety
        Criteria of General Services Agency, General Service
        Administration, Washington, DC

/22/   BURROS, R.H.; Probability of Failure of Buildings from
        Fire, J. Structural Division, ASCE, 101 (1975) 1947-60

/23/   WILLIAMSON, R.B., LING, W.T.; The Use of Probabilistic
        Networks for Analysis of Smoke Spread and Egress of
        People; First International Symposium on Fire Safety
        Science, Gaithersburg, ML, 1985.

HIGH TEMPERATURE EFFECTS

J A Purkiss

Department of Civil Engineering & Construction
Aston University
Birmingham B4 7ET, UK.

ABSTRACT

The paper reviews the need for data on material properties at high temperatures and then considers, both for steel and concrete, the thermal and mechanical behaviour. With respect to mechanical behaviour the concept of the total strain model is discussed and suggestions made on the need to standardize testing methods.

INTRODUCTION

This paper does not attempt to provide data illustrating the thermal or mechanical behaviour of construction materials at elevated temperatures (this being the remit of subsequent papers), but it is intended to provide a background to the reasons and methods of carrying out tests to provide these data and to examine the use and validity of the results of such tests.

Material behaviour may be divided into two categories – thermal and mechanical (1,2,3). It is essentially the thermal properties, with the exception of thermal expansion, that are required for evaluation of the serviceability limit state (i.e. temperature rise on an unexposed face) and the mechanical properties that are required to evaluate the ultimate limit state of structural integrity. Note that even here the thermal properties are required to enable a prediction to be made of the temperature regime within a structure or structural element during a fire.

Table One provides a check list of properties which need evaluating for differing construction materials (1,2,3). However, this paper will be, in general, limited to discussing the two major construction materials – concrete and steel.

Table 1 Properties of Materials Considered to be of Interest

Property or characteristic	Concrete	Steel	Masonry	Wood	Plastics	Gypsum
PHYSICAL						
Density	X	X	X	X	X	X
Expansion	X	X	X	–	X	X
Softening	–	X	–	–	X	–
Melting	–	–	–	–	X	–
Spalling	X	–	–	–	–	–
CHEMICAL						
Decomposition	–	–	–	X	X	–
Charring	–	–	–	X	X	–
MECHANICAL						
Tensile strength	X	X	–	X	X	X
Compressive strength	X	X	X	X	X	X
Modulus of elasticity	X	X	X	X	X	–
Strain/stress relationships	X	X	–	–	X	–
Creep	X	X	–	–	X	–
THERMAL						
Conductivity	X	X	X	X	X	X
Specific heat	X	X	X	X	X	X

Note – (x) applies, (–) does not apply.

THERMAL PROPERTIES

In essence the evaluation of properties within this category is a matter of standard routine testing with the exception of thermal strain (or thermal expansion) which will be considered in this paper as a mechanical property and on which discussion will be delayed until a subsequent section.

In order to evaluate the thermal response of a structural element it is necessary to have data on the variation of any three of specific heat (c), density (ρ), thermal conductivity (k), and thermal diffusivity (ψ), since

$$\psi = k(\rho\, c) \qquad (1)$$

For structural or reinforcing steel there is a little difficulty in establishing values but the problem with concrete, and, to a lesser extent, masonry, is the presence at temperatures below about 150°C of pore water which will significantly effect the values of thermal conductivity and specific heat (hence also thermal diffusivity). It thus becomes essential that the initial moisture content be stated for all test results. It should be noted that any physical/chemical changes in the aggregate will also effect results.

Knowledge of the specific heat etc. will only allow the solution of the heat diffusion equation,

$$\mathrm{div}\,(\psi\,\mathrm{grad}\,T) = \dot{T} \qquad (2)$$

to be used to determine the temperature (T) within the structural element. It is still necessary for the boundary conditions at the fire structure interface to be evaluated. The mechanisms of heat transfer by both convection and radiation must be included. Most of the heat transfer occurs by radiation. It is possible to adjust the values of view factors and emissivites to give a reasonable fit to temperature profiles from predictions compared with standard furnace tests. Although guidance is available on the values of these parameters (4), it has been pointed out (5) that even a slight variation in these parameters can cause a substantial change in predicted temperatures. Thus it is essential to be able to evaluate these parameters, if at all possible, from first principles. Some further guidance is given in a recent reference by Drysdale (6).

It is also convenient within this section to consider the explosive spalling of concrete although it will be dependant on both the stress levels within the concrete and the strength of the concrete. In a recent report by Malhotra (7) it becomes clear that mechanism of spalling is not understood although the factors influencing its seemingly random occurence can be delineated. It is also possible to produce advice on methods to attempt to alleviate its occurence and severity.

MECHANICAL PROPERTIES

The aim behind a study of the mechanical properties of a material is to enable a constitutive law governing the bahaviour of that material to be established. This section of the paper will initially only consider a uniaxial state of stress (or strain) when it may be considered that such a constitutive law takes the following form,

$$F(\sigma, \tilde{\sigma}, \varepsilon, T, \tilde{T}, t) = 0 \qquad (3)$$

where F is a generalized function, σ the current stress state, $\tilde{\sigma}$ the stress history, ε the current strain level, T the current temperature, $\tilde{T}$ the temperature history and to the current time (either real or parametric). Currently no such generalized function, F, exists and recourse must be had to a simplified model such as the total strain model proposed by Anderberg and Thelandersson (8) where the total strain (ε_T) is decomposed into four components:-

    i)   thermal strain (ε_{th})
    ii) stress related strain (ε_{σ})
    iii) creep strain (ε_{cr})
    iv) transient strain (ε_{tr})

such that

$$\varepsilon_T = \varepsilon_{th}(T) + \varepsilon_{\sigma}(\sigma, \tilde{\sigma}, T, \tilde{T}) + \varepsilon_{cr}(T, \tilde{T}, \sigma, \tilde{\sigma}, t) + \varepsilon_{tr}(\sigma, \tilde{\sigma}, T, \tilde{T}) \qquad (4)$$

It is recognized (1,2,3) that a series of differing testing regimes is required to evaluate Equations 3 or 4. These are summarized in Fig. 1. The recognized existence of these testing regimes does not, currently at least, imply the existence of standardized testing methods and it should be noted that substantial variations will be caused by variations in rates of straining (or loading), rate of temperature rise, specimen size and initial moisture content, besides those additionally due to aggregate type and mix design, for concrete, and the type of steel, for structural steelwork or reinforcement. It

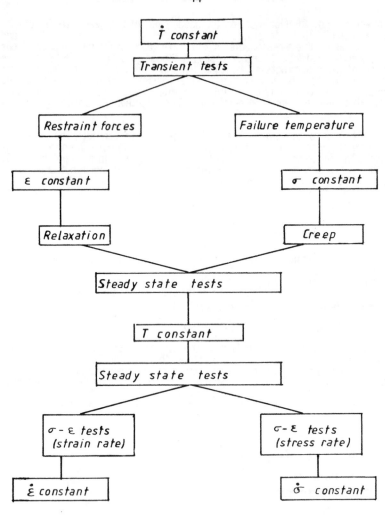

Fig 1   Testing régimes for mechanical properties

thus becomes apparent that care is needed in comparing test data from differing sources.

For concretes, the rate of temperature rise is important since this will cause the existence within the specimen of a set of self-equilibrating parasitic stress that will during any heating or cooling cycle affect the results obtained on that specimen. The temperature gradient within the specimen can be reduced by either using small diameter specimens which will only allow the use of small aggregate sizes or by using a low heating rate in the furnace which may give rise to problems with design or control of the furnace. Inevitably, a compromise solution is used by limiting the aggregate size to 10 mm and the heating rate to around 2 to 5 deg C/min.

Before discussing the validity of the total strain model, it is convenient to discuss the evaluation of each of the constituent terms for both concrete and steel. For timber this is not needed as the performance of a timber element is based on "dry" stresses on the remaining portion of the member not removed by charring.

a) CONCRETE

The following discussion has been written in terms of evaluating the compressive behaviour of concrete. A similar procedure could be carried out for concrete in tension, except there exist very few data. It is perhaps fortunate that the tensile behaviour of concrete is not very important in determining structural response.

i) Thermal Strain (ε_{th})

This is the unrestrained thermal expansion after allowance has been made for drying shrinkage, although it is often more convenient to use a composite model, for the purpose of calculating fire response, which includes drying shrinkage. During heating parasitic self-equilibrating stresses will be locked in due to the differing magnitudes of natural expansion of both the mortar matrix and the aggregate. These stresses will depend on the thermal expansions of the components, values of the Young's Moduli, Poisson's ratio of the matrix and the aggregate, and the volume concentration of the aggregate. Dougill (9) has given some indication of the importance of these parasitic stresses, and has also shown that it is possible to predict the value of the overall thermal strain of the concrete given the basic properties of the aggregate and the cement paste.

ii) Stress related strain (ε_{σ})

Strictly this should be the instantaneous stress related strain, i.e. the tests should be performed under extremely high strain (or loading) rates. However, most tests are carried out at finite, often low, rates of strain (or loading), thus an indeterminate element of creep will be present in results and thus, most certainly at fairly high temperatures, creep strains may be partially evaluated twice.

Tests may either be carried out under load or strain control. If the former be used then it will not normally be possible to determine the post

46

peak behaviour. If the latter, then the whole characteristic may be evaluated. The post peak behaviour is important for determining accurate structure response, as has been demonstrated by Dougill (10, 11). The post peak behaviour (i.e. slope of the descending branch) is temperature dependant (12) and tends to flatten with increase in temperature. Very little work has been done on unload and reload cycles for concrete at elevated temperatures, although that little which has, seems to suggest that concrete obeys Prandtl's rules; and seems to unload and reload along a line parallel to the initial tangent. It is by no means certain what occurs if a temperature change is imposed during an unloading-reloading cycle, or if the concrete unloads as far as to go into a state of tension.

Most tests have been performed on specimens heated without load, although as early as 1956 Malhotra (13) showed that the strength reduction is less if the concrete is heated with a preload than without. The test with no preload, does in any case appear to be meaningless as far as structural response is concerned, since any member in a fire will, at least, be subjected to stresses imposed due to the dead weight of the structure before commencement of the fire. It has been shown (14, 15, 16) that the reduction of peak stress is less for concretes heated under preload, as is the reduction in the initial tangent modulus. These results also seem to indicate that the strain at peak stress remains sensibly constant over a substantial temperature range. A corollary to this is that there may not for regions subjected to high temperatures and compressive stresses be sufficient ductility present to allow the formation of full plastic hinges and thus invalidate a basic assumption inherent in the use of plastic collapse methods.

iii) Creep strain (ε_{cr})

Tests are generally performed by heating the specimen with no load to the test temperature, allowing the test temperature to stabilize then applying the test load and measuring the resultant strains over a time period. To enable structural performance in a fire to be calculated a time period of 5 hours is sufficient. During this time primary creep and some secondary creep will occur. It is possible that at very high temperature and stress levels creep rupture may be observed.

The method described above is simple but has two drawbacks. The first is that the test load should be applied instantaneously, otherwise some creep will occur during the loading period. The second is that in a structure concrete will be heated under load. Creep tests performed on specimens heated under load generate lower creep strains (13), since in standard creep tests the parasitic stresses built up during heating need to be overcome initially.

Various theories have been postulated for creep behaviour at ambient conditions. At elevated temperatures it is generally agreed that an Ahrenius type rate reaction occurs (16,17,18) and that a power law for time lapse with an exponent of a round 0.5 is valid (8,16,19). Although it is frequently assumed that the concept of specific creep is valid, this is by no means certain as even at low stress levels creep may be non-linear with respect to the applied stress level (20).

iv) Transient strain ($\mathcal{E}_{tr}$)

Until a decade, or so, ago the majority of investigators considered only the need for steady state tests. It was not until Thelandersson and Anderberg (8) performed tests under constant stress and varying temperature, measuring the resultant strain was it realized that transient effects were important. The transient strain is in part due to aggregate-mortar incompatibilities and in part due to problems in the application of instantaneous loading during stress-strain tests and creep tests and in part due to the imposition of thermal gradients during heating in an all types of tests. It is perhaps significant to point out that for steel where none of the above difficulties arise there is no element of transient strain. It has been pointed out by Dougill (21) that care is needed in applying the concept of transient strain to high heating rates, since most tests to determine transient strain have been done at relatively low heating rates (e.g. 5 deg C/min).

Transient strain cannot itself be measured but must be determined from the remaining four components:

$$\mathcal{E}_{tr} = \mathcal{E}_T - \mathcal{E}_\sigma - \mathcal{E}_{cr} - \mathcal{E}_{th} \tag{5}$$

For siliceons aggregate concrete it has been found (8,14) that at temperatures below 550°C (which is the temperature at which the quartz phase change occurs),

$$\mathcal{E}_{tr} = -k \, \mathcal{E}_{th} \, \beta \tag{6}$$

where k is a constant of order 2, although its exact magnitude will almost certainly depend on the aggregate-cement ratio of the concrete, and β is the ratio of the applied stress to the ambient cylinder strength. Above 550°C the following equation holds,

$$d\mathcal{E}_{tr} / dT = 10^{-4} \beta \tag{7}$$

where d $\mathcal{E}_{tr}$/dT is the rate of change of transient strain with respect to temperature.

v) Total Strain ($\mathcal{E}_T$)

This is evaluated by preloading a specimen to the test load, then increasing the temperature (of the furnace) at a known rate, whilst keeping the load constant, until such time as the specimen can no longer support the applied load. During the test measurements of the total strain are taken.

To date, validation of the total strain model only appears to have been undertaken up to applied stress levels of 0.45 σ_{max} (where σ_{max} is the ambient cylinder strength). It is not clear how accurate the model is if, say, a stress of 0.8σ_{max} be applied or if a point on the descending branch of the stress-strain curve be used.

More recently Thelandersson has extended the total strain model to multi-axial stress states (22), but it should be pointed out that stresses due to thermal incompatibilities become even more important in a multi-axial stress state and may lead to premature failure (9).

The little work that has been carried out on tensile strength behaviour of concrete at elevated temperatures (8,23,24,25) indicates that the deterioration in strength is greater than that in compression.

b) STEEL

For normal steel sections, the thickness is such that the temperature in the section may be considered uniform and thus the problem met with concrete on heating rates is nowhere near as important.

Also, all the stress related tests have been carried out in tension, and it is then assumed that the behaviour in compression is similar provided any tendency for lateral instability in the steel is prohibited.

i) Thermal strain (ε_{th})

This is a straightforward evaluation of the thermal expansion.

ii) Stress related strain (ε_σ)

Similar comments to those on concrete concerning straining (or loading rates) apply.

iii) Creep strain (ε_{cr})

An extension of the Dorn temperature compensated time theory has been used (26,27,28). From such data that are available, it is seen that the parameters involved are highly sensitive to the steel type and thus creep data must be treated with extreme care.

Also under this heading, for pre-stressing steel stress relaxation needs to be considered.

iv) Transient strain (ε_{tr})

For steel the transient strain component does not exist.

It appears that the total strain model is a reasonable representation for steel in tension, and thus by implication also in compression.

USE OF THE STRESS-STRAIN MODEL

a) Concrete

Where flexural behaviour is dominant and the concrete compression face is remote from the fire e.g. at midspan of a beam or slab, knowledge of the "elastic" stress-strain relationship (or maximum stress) can be sufficient to determine the collapse load. This proceedure will certainly be sufficient for an under-reinforced slab. It should produce a safe solution at the supports of such a beam, thus underestimating the time to collapse. The situation is not simple when the concrete is in heavy compression as in a column, and there is as yet no simple solution to this problem. For complete structures a full constitutive law will be necessary.

b) Steelwork

Currently calculations are performed using only a reduced yield stress (and elastic modulus) since except at very high temperatures (i.e. a low state significant of applied load) creep may not be sufficient. Only in more detailed analyses of complete structures will a full constitutive law be necessary.

STANDARDIZATION OF TEST METHOD

It becomes obvious from the above discussion that although the testing regimes have become identified and standardized the test method has not. As a result of the deliberation by RILEM on material behaviour, a committee has also been set up to study test methods.

The main problems are standardization of straining (or loading) rates and heating rate.

i) Straining rate

To obtain the instantaneous stress-strain response the strain rate must must be set as high as possible to reduce the amount of creep. It is, however, inevitable that a compromise solution will be used owing to the problems of measuring stresses and strains at high straining rates and mechanical problems of the rig itself.

ii) Loading rate

Since a load controlled test is usually incapable of determining the whole stress-strain characteristic, load controlled tests ought not to be considered.

iii) Heating rate

Since in an actual structure the heating rate for a particular portion will not be constant over a time period and since the effect of differing temperature rates on the mechanical properties of materials has not been completely studied, mechanical properties should be evaluated at a series of different rates of temperature rise.

The common way of controlling temperature rise is to allow the furnace temperature to rise at a constant rate. However, this will not, due to thermal gradients within the specimen, give a constant rate of temperature rise to the specimen. It could well be argued that the important variable is that of temperature rise within the specimen, which would, however, be difficult to control and thus yet again a compromise solution would need to be reached.

CONCLUDING REMARKS

It is recognised that in spite of criticisms of testing methods and comparibility of results much valuable work has been done in a relatively short time, which is enabling the engineer to turn toward designing a structure for fire with a similar degree of certainty as against any other load system. However, it must also be added that the remains much still to be done, especially in standardizing testing and obtaining mathematically sound models for material behaviour at elevated temperatures.

REFERENCES

1 Schneider, U., (Ed), Properties of Materials at High Temperatures - Concrete. Dept of Civil Engineering, Gesamthochschule Kassel, 1981.

2 Anderberg, Y., (Ed), Properties of Materials at High Temperatures - Steel, Sweden, Lund Institute of Technology, 1985.

3 Malhotra, H.L., Design of Fire Resisting Structures, England, Surrey University Press, 1982.

4 Pettersson, O.; Magnusson, S.E.; Thor, J. Fire Engineering Design of Steel Structures, Publication No 50, Swedish Institute of Steel Construction, Stockholm, 1976.

5 Drysdale, D.; An Introduction to Fire Dynamics, Chichester, John Wiley & Sons, 1985.

6 Weeks, N.J.; The Lateral Instability of Slender Reinforced Concrete Columns Subject to Fire, Ph D Thesis, Aston University, 1985.

7 Malhotra, H.L.; Spalling of Concrete in Fires, CIRIA Technical Note No 118, London, 1985.

8 Anderberg, Y., Thelandersson, N.; Stress and Deformation Characteristics of Concrete at High Temperatures, Bulletin 54, Lund Institute of Technology, Lund, 1976.

9 Dougill, J.W.; Some Effects of Thermal Volume Changes on the Properties and Behaviour of Concrete, International Conference of the Structure of Concrete, University of Southampton, Willey, 1965.

10 Dougill, J.W.; Modes of Failure of Concrete Panels Exposed to High Temperatures, Mag. Conc. Res. 24 (1982), 71-76.

11 Dougill, J.W.; Conditions for Instability in Restrained Concrete Panels Exposed to Fire, Mag. Conc. Res. 24 (1982), 139-148.

12 Furamura, F.; The Stress-strain Curve of Concrete at High Temperatures, Annual Meeting of the Architectural Institute of Japan, 1966.

13   Malhotra, H.L.; The Effect of Temperature on the Compressive Strength of Concrete. Mag. Conc. Res. 8 (1956), 85-94.

14   Schneider, U.; Behaviour of Concrete under Thermal Steady-state and Non-steady State Conditions, Fire and Materials, 1 (1976), 103-115.

15   Fischer, R.; Über das Verhalten von Zementmortel und Beton bei höheren Temperaturen, Deutscher Auschuss für Stahlbeton, 214 (1970), 60-218.

16   Bali, A.; The Transient Behaviour of Plain Concrete at Elevated Temperatures, Ph D Thesis, Aston University, 1984.

17   Maréchal, J.C.; Fluage du Béton en Fonction de la Température, Annls. Inst. Tech. Bâtim, 23 (1970), 13-24.

18   Maréchal, J.C.; Fluage du Béton en Fonction de la Température - Compléments Experimentaux, Matériaux et Construction, 3 (1970), 395-406.

19   Gillen, M.; Short Term Creep of Concrete at Elevated Temperatures, Fire and Materials, 5 (1981), 142-8.

20   Frendental, A.M., Roll. F.; Creep and Creep Recovery of Concrete under High Compressive Stresses, J. Am. Concr. Inst. 29 (1958), 1111-1142.

21   Dougill, J.W.; Materials Dominated Aspects for Structural Fire Resistance of Concrete Structures, Fire Safety of Concrete Structures, S. Abrams (Ed), ACI SP 80, Detroit, 1983, pp 151-175.

22   Thelandersson, S.; On the multi-axial behaviour of concrete exposed to high temperature, Nucl. Engng. London, 75 (1982), 271-282.

23   Zoldners, N.G. Effect of High Temperatures on Concretes incorporating different Aggregates, Proc. Am. Soc. Test, 60 (1960), 1087-1108.

24   Purkiss, J.A.; Steel Fibre Reinforced Concrete at Elevated Temperatures, International Journal of Cement Composites & Lightweight Concrete, 6 (1984), 179-184.

25   Purkiss, J.A. Some Mechanical Properties of Glass Reinforced Concrete at Elevated Temperatures, 3rd International Conference on Composite Structures, Paisley, 1985.

26   Harmathy, T.Z. A Comprehensive Creep Model, J. bas. Engng. 89 (1967), 496-502.

27   Harmathy, T.Z., Stanzack, W.W. Elevated Temperature Tensile and Creep Properties of some Structural and Prestressing Steels, National Research Council of Canada, Division of Building Research, Paper No 424, Ottawa, 1980.

27   Anderberg, Y. Armeringstähls Mekmasken Egenskaper vid Höga Temperatur, Bulletin No 61, Lund Institute of Technology, Lund, 1978.

MODELLING OF CONCRETE BEHAVIOUR AT HIGH TEMPERATURES

SCHNEIDER, ULRICH   PROF.DR.-ING.

Department of Civil Engineering, University of Kassel,
Mönchebergstr. 7, D-3500 Kassel, Germany

ABSTRACT

    The use made of material properties in theoretical studies may be
different. It depends on the individual approach, the objective and the
quality of result required. This report gives a condensed survey of the
present state of knowledge in the field of high temperature properties of
concrete, which may assist in giving an answer to the problem of estimating
the fire behaviour of concrete members. A new material model based on
recent research results in being developed and discussed.

The report is divided into three different parts. Chapter 2 contains a
brief introduction to various concrete test methods. Chapter 3 comprises
the properties of concrete according to the existing literature and in
chapter 4 a description of analytical models for the calculation of fire
behaviour of concrete elements is given.

1. INTRODUCTION

    Many investigations on the effect of fire on concrete and concrete
elements have been reported during the last four decades /1 to 4/. Nevertheless the thermal and mechanical properties of concrete at high temperatures are still under consideration. This is due to the fact that many of
the reported test results are hard to interpret as the

- tested types of concretes are different and the descriptions of the
  tests are incomplete,

- employed test procedures are different and the test conditions are not
  comparable,

- shapes of the test specimens and the test apparatus used are different.

In this report the main features of concrete behaviour at high temperatures are summarized and discussed. It is essentially a survey of existing
data with respect to real fire situations i.e. a temperature range from
20 to 1.000°C is considered whereby the material behaviour under short-
time exposure periods from 0.5 to 5 hours is discussed preponderantly.

In the paper the following particular properties are identified:

- thermal properties
- compressive strength
- modulus of elasticity
- stress and strain characteristics
- thermal strain and shrinkage
- transient state creep and restraint.

As the fire exposure normally leads to a comparatively rapid loss of moisture in the concrete elements the properties mentioned above are usually determined with unsealed concrete specimens. Testing of unsealed concrete probes means that all properties are more or less influenced by the rapid drying of the cement gel. In so far one should strictly speak rather of drying creep and dry compressive strength than purely of creep and strength as stated above. For the sake of simplicity the term drying will be omitted in the following. But one has still to keep in mind that the drying process is one of the main effects determining the concrete behaviour under thermal exposure.

2. DETERMINATION OF CONCRETE PROPERTIES BY DIFFERENT TEST METHODS

The concrete properties are closely related to the specific test method employed /3/. Therefore, the possibility of conversion of such properties into material equations depends on many factors. If mechanical properties are considered, adequate theological models must be developed. In many cases this is not possible, rherefore test methods, which are closely related to practical conditions, are to be preferred. Under fire conditions the concrete is subjected to transient processes and therefore there is an urgent need for the measurement of those properties, which are determined under transient conditions. These properties should be distinguished from other properties derived under steady state conditions.

The three main test parameters are heating, the application of the load and the control of strain. These can have fixed i.e. constant values or be varied during testing giving transient conditions. Six practical regimes, which can be used for determining mechanical properties, are illustrated in fig. 1 and described below.

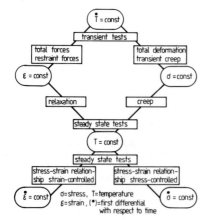

Fig. 1

Different testing regimes for determining mechanical properties of concrete at high temperatures

Steady state tests

I.   Stress-strain relationships, stress rate controlled:
     During the test the specimen is heated to the desired test tempera-
     ture $(T_1, T_2$ or $T_n)$ starting at time $t = 0$ with a constant heating rate
     $\dot{T}$. For practical reasons (size of the specimen) and taking into account
     exposure in fire conditions, the heating rate should be in the range
     of 0,1 K/min to 10 K/min. After an initial preheating period at time t^*
     say, the specimen is subjected to a constant rate of loading $\dot{\sigma}$ = const.
     From the test data the modulus of elasticity, the compressive strength
     and the ultimate strain can be determined.

II. Stress-strain relationship, strain rate controlled:
     This test method is closely related to the stress controlled σ-ε-tests
     previously discussed. The specimen is heated to the required tempera-
     ture T^* at a constant heating rate $\dot{T}$. After the initial heating period
     when the specimen has reached constant temperature it is loaded at a
     constant strain rate $\dot{\varepsilon}$. This procedure yields complete σ-ε-curves and
     from this it is possible to determine the maximum mechanical energy
     the specimen dissipates during fracture (ultimate mechanical dissipa-
     tion energy).

III. Creep:
     In steady state creep tests the specimen is (slowly) heated to the
     desired temperature T^*. When thermal equilibrium is reached $(t = t^*)$
     the load is applied. The temperature T^* and the load σ_o applied are
     kept constant during the whole test period. At time t^*, when the load
     is applied, an instantaneous elastic deformation occurs and thereafter
     under sustained constant load creep deformation takes place. The test
     has little relevance with respect to the fire situation as the test
     periods are normally far beyond the duration of building fires.

IV. Relaxation:
     Initially a heating procedure analogous to the steady state creep test
     is applied. At time t^* when thermal equilibrium is reached the load is
     applied and the instantaneous elastic strain recorded. The initial
     strain is kept constant during the whole test period and the stress
     level is being recorded. The test has little relevance with respect to
     the fire situation as the test periods are normally far beyond the
     duration of building fires.

As the report is devoted to the modelling of concrete under fire the results of steady state creep and relaxation tests will not be discussed
further on.

Transient tests

V.   Total deformation and transient creep:
     At time $t = 0$ the specimen is subjected to a certain constant applied
     load σ_o. Thereafter it is subjected to a constant heating rate. For
     practical reasons (size of the specimen) and taking into account expo-
     sure to fire conditions the heating rate should be in the range of
     0,1 K/min to 10 K/min. Heating is continued until failure occurs.
     During the whole test period the strains ε_{tot} of the specimen are re-
     corded. If $\sigma_o \equiv 0$ pure thermal expansion occurs.

VI. Total forces and restraint:
The concrete specimen is loaded at time t = 0 up to a given constant load level σ. The initial compressive strain is recorded. Thereafter it is subjected to a constant heating rate. The initial elastic strain is kept constant during the whole heating period by varying continously the applied external load. Heating is usually continued until the measured total force drops below its original load level at time t = 0.

3. CONCRETE PROPERTIES AT HIGH TEMPERATURES

3.1 Thermal Properties

Concrete properties which are necessary to calculate the heat transfer and temperature distributions in concrete members are the specific density ρ, the thermal conductivity k, the heat capacity c_p and derived from those the thermal diffusivity a. The density of concrete shows only a little temperature dependance as indicated in fig. 2 which is mostly due to moisture losses during heating. However limestone concretes show a significant decrease of density at about 800°C due to the decomposition of the calcareous aggregate.

The thermal conductivity of concrete depends on the conductivities of its constituents. The major factors are the moisture content, the type of aggregate and the mix proportions. The conductivity of any given concrete varies approximately linear with the moisture content. Up to 100°C the conductivity seems to increase with temperatures. Thereafter a loss of conductivity is observed. With lightweight concretes the conductivity may be nearly constant or slightly increasing up to temperatures of 1.000°C. In this case the density of the aggregate is the decisive parameter. Fig. 3 shows some typical data for normal and lightweight concretes.

In the context of thermal properties the specific heat seems to be that property which is least understood. The specific heat at constant pressure is defined as

$$c_p = (\frac{\partial H}{\partial T})_p \tag{3.1}$$

where H = enthalpy, T = temperature and p = pressure. In technical reports normally average values of $\bar{c}_p$ are used. They are defined by the equation:

$$\bar{c}_p(T) \cdot (T-T_o) = \int_{T_o}^{T} c_p(T) \, dT \tag{3.2}$$

If the heating of the material is accompanied by chemical reactions, the enthalpy is a function of the degree of conversion from the reactants into the products ζ, $0 \leq \zeta \leq 1$, as well as of temperature. Thus the first equation becomes:

$$c_p = (\frac{\partial H}{\partial T})_{p,\zeta} + (\frac{\partial H}{\partial \zeta})_{p,T} \cdot \frac{d\zeta}{dT} \tag{3.3}$$

where the second term can clearly be recognized as the latent heat and the first term as the sensitive heat contribution to the specific heat. According to Dulong-Petit's rule no considerable spread of c_p of different concretes is to be expected. Differences may be caused by the latent heat of

the different reactions during heating (water release, dehydration, de-
carbonization, α→β quartz inversion). From the reported test results it
can be stated (see fig. 4):

- The type of aggregate indicates no clear effect on the heat capacity if
  temperatures below 800°C are considered. Where temperatures of 800°C are
  exceeded with calcareous concrete c_p rises immediately due to decarbo-
  nization.

- The mix proportions influence the heat capacity in so far as richer
  mixes indicate a higher latent heat due to dehydration effects.

- The water content is important at temperatures below 200°C. Wet concretes show an apparent specific heat nearly twice as high as oven-dried
  concretes.

The thermal diffusivity of concrete is determined by the thermal properties of its constituents or it may be evaluated from non-steady state
measurements. The variations of the reported data are comparatively high
and may be attributed to the type of test method, the type of concrete
under consideration or to the specific treatment of the specimens prior to
the tests. As the diffusivity values are decisive for the temperature calculations in concrete elements under fire,there is an urgent need for most
reliable and accurate thermal data. Fig. 5 contains some results of measurements and a proposed empirical correlation for a structural concrete
with quartzite aggregate.

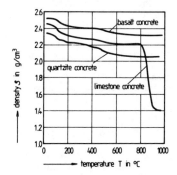

Fig. 2: Density of structural
      concretes at high tempera-
      tures

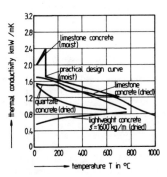

Fig. 3: Thermal conductivity of
      different structural con-
      cretes

In this connection it should be noted that the accuracy of the nonsteady
test methods is estimated to be usually ± 10 percent with respect to the
thermal diffusivity and ± 15 percent with respect to the thermal conductivity.

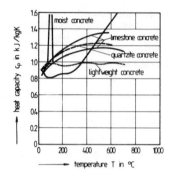

Fig. 4: Heat capacity of different   Fig. 5: Thermal diffusivity of
        concretes   different concretes

3.2 Compressive Strength

The compressive strength of different concretes has been the subject
of many investigations. Some of the main findings are summarized in fig.6.
It should be noted, that the tests under discussion were generally performed with unsealed concrete specimens. In most cases the loading rates
have not been stated, however, it seems that within a normal range the
loading rates have neglectible influence on the high temperature strengths.
From the reported data the following general conclusions can be drawn:

- Original strength and water-cement ratio within the practical range of
  usage for concretes hardly influence the high temperature-strength characteristics.
- Aggregate-cement ratio has a significant effect on the strength on concrete exposed to high temperatures. The reduction being proportionally
  smaller for lean mixes than for rich mixes.
- Different types of aggregates influence the strength-temperature characteristics. The decrease in strength of calcareous and lightweight
  aggregate concretes occurs at higher temperatures compared to siliceous
  concretes.
- Type of cement has little effect on strength-temperature characteristics.
- Maximum size on aggregate seems to be a second order factor as investigations of mortars and various concretes demonstrate.
- Sustained stresses during the heating period influence the shape of the
  strength-temperature relationship significantly. It is evident, that the
  "stressed strength" is higher than "unstressed strength". The stress
  level itself hat little effect on the ultimate strength as long as $\alpha >$
  0,20 but becomes important if $\alpha < 0,20$.
- Rate of heating has little effect as long as temperature gradients in the
  test specimens are limited (< 10°C/cm).
- Residual compressive strength values are lower than the belonging high
  temperature strength values.

Often the evaluation of concrete structures requires data which enable the
determination of multiaxial states of stresses. Especially if plates or
slabs are to be considered the application of uniaxial material properties
may lead to unexpected errors or incorrect results. The biaxial high tem-

perature strength of concrete has been studied during the recent two years /5/. Fig. 7 shows the failure envelop of a structural concrete under biaxial compression. It is clearly indicated that the biaxial compressive strength is higher than the uniaxial strength irrespective of the individual stress ratio and temperature level. Similar results have been obtained with a structural lightweight concrete (fig. 8). Further it was noted that the relative increase of strength at high temperatures under biaxial state of stresses is significant. Especially at temperatures above 450°C - the effect of biaxial stresses indicates its increasing importance with respect to the material failure.

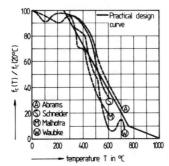

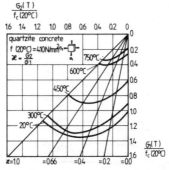

Fig. 6: High temperature compressive strength of ordinary structural concrete

Fig. 7: Biaxial compressive strength of normal concrete at high temperatures (after Ehm)

3.3 Modulus of Elasticity

A limited number of publications consider the elastic properties of concrete at high temperatures. Essential results are summarized in fig. 9. From the presented data it can be stated that:

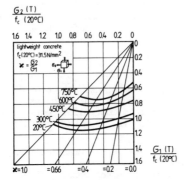

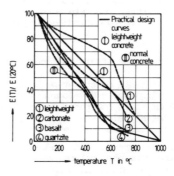

Fig. 8: Biaxial compressive strength of lightweight concrete at high temperatures (after Ehm)

Fig. 9: Modules of elasticity of structural concrete with different types of aggregate

- Original strength of concrete and water-cement ratio seem to have little influence on the elasticity-temperature relationship.
- Type of aggregate has in most cases a strong influence. Lightweight aggregate concretes indicate the lowest decrease of the modulus of elasticity and siliceous aggregate concretes the highest one. The range of data from different workers varies significantly (comp. fig. 9).
- Type of cement has little effect on modulus of elasticity-temperature characteristics.
- Sustained stresses during heating of the test specimen significantly affect the elasticity-temperature behaviour. "Stressed elasticities" are always higher than "unstressed elasticities". The stress level itself has little effect within a range of α = 0,1 to 0,3 (comp. fig. 10).

Data on the Poisson's ratio μ at high temperatures are rare. Fig. 11 shows the results of Ehm /5/ derived in 1985. At 20°C the Poisson's ratio is constant until the load level exceeds 70% of the ultimate load. With increasing temperatures a significant deviation of μ from its original values occurs. In some cases μ > 0.5 was observed. These values indicate material effects which occur far beyond the elastic range.

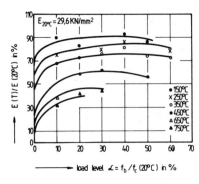
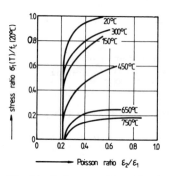

Fig. 10: Modulus of elasticity of normal concrete being stressed during heating up

Fig. 11: Poisson ratio of uniaxially loaded concrete at high temperatures (after Ehm)

3.4 Stress and Strain Characteristics

Most tests of this type have been performed by stress-rate controlled tests. With recent developments in the field of modern hydraulic test equipment it is now possible to perform strain-rate controlled tests, too. Those tests yield σ-ε-curves as indicated in fig. 12. From the figure and the literature reviewed the following conclusions can be drawn:

- Original strength and the water-cement ratio within the practical range of concrete application hardly influence the shape of σ-ε-curves.
- Aggregate-cement ratio has a significant effect on the modulus of elasticity and consequently also on the initial slope of the σ-ε-curves. Mortars (high cement content!) indicate a lower initial slope than normal concretes. σ-ε-curves of concrete indicate a somewhat greater curvature than those of mortars.

- Type of aggregate is the main factor affecting the shape of the σ-ε-curves. Concretes made with hard aggregates (siliceous, basaltic) generally have a steeper decrease of the initial slope with increasing test temperatures than those with softer aggregates (e.g. lightweight aggregates).
- Lightweight concretes indicate practically only little changes in the shape of the σ-ε-curves for temperatures up to about 250°C.
- The ultimate strain (strain at the failure point or maximum strength of a σ-ε-curve) turned out to be nearly independent of the type of aggregate.
- The temperature dependant dissipation energy (i.e. work of fracture in a strain-rate controlled compressive test) indicates a maximum in the temperature region of 300 to 600°C. The value of the ultimate dissipation energy for the low and high temperature regions lies between 20 and 70 J/kg. Maximum values from 60 to 100 J/kg have been found in the medium temperature region from 300 to 600°C.
- Type of cement seems to be of minor influence as fas as concretes are considered. Mortars (mix proportion 1:3:0,5) made with different types of cement showed significant differences.
- Curing conditions influence the stress-strain behaviour at relative low temperatures (< 300°C). Usually the initial slopes of the σ-ε-curves and stress maxima are lower for specimens cured under water than for dried or air cured specimens. However, in most cases the σ-ε-relationships are normalized to the ultimate strength at 20°C ($f_c(20°C)=1$)). Sometimes by this an apparent rise of the high temperature values of the compressive strength occurs if water cured specimens data are taken as reference values.
- A sustained load (prestess e.g.) during heating varies the shape of the σ-ε-curve significantly which is independent of the type of concrete being tested. Specimens under a sustained load (load level α) during the heating period indicate a significant relative increase of compressive strength and modulus of elasticity compared to specimens which were not loaded during heating but tested under the same conditions. The ultimate strain is also significantly reduced with loaded specimens (comp.fig.13). Up to test temperatures of about 450°C concrete specimens indicated similar behaviour to unheated specimens. The σ-ε-curves are nearly independant of the test temperatures. The load level during heating itself seems to have a minor influence.
- Stress-strain relationships under biaxial conditions have been derived by Ehm /5/. Fig. 14 shows a typical test result for normal concrete tested with a stress ratio ℒ = 1:1 i.e. during the tests simultaneous stress increase of the same order in both axis was performed. The observed ultimate strains are somewhat higher than in the uniaxial case. Especially in the temperature range above 450°C a significant increase in plasticity occurs.

3.5 Thermal Strain and Shrinkage

The objectives of thermal expansion measurements concern the question of how the thermal strain is effected by single components of the composite material and how the incompatibilities of the thermal strains of the composites influence the mechanical properties of concrete at high temperatures. Some test results of thermal strain measurements are reviewed in fig. 15. From the figure it can be stated that:

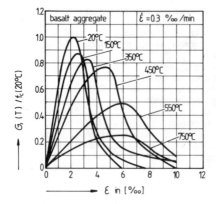

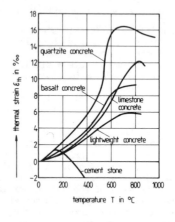

Fig.12: Stress-strain relationship
for normal concrete derived
in strainrate controlled tests

Fig.13: Ultimate strain ε_{ult} as a
function of temperature for
specimens stressed at
different levels during
heating up (σ-ε-tests)

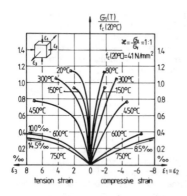

Fig.14: Biaxial stress-strain relationship of normal concrete
under a stress ratio $\mathfrak{X}$=1:1
(after Ehm)

Fig.15: Thermal strain of different
concretes

- Thermal strain is a non-linear function of temperature, even at relatively
  low temperatures.
- The main factor affecting the thermal strain is the type of aggregate,
  the coarse aggregate fraction plays a dominant role.

- Pure hydrated cement paste indicates contraction shrinkage at temperatures above 150-400°C.
- At very high temperatures (600-800°C) most concretes indicate no or a reduced expansion. In some cases the concrete shrinks due to chemical or physical reactions in the aggregates.
- Thermal strain measurements normally include shrinkage. This is unavoidable as the tests are performed with unsealed specimens.

Compared to the thermal expansion shrinkage strains are small. Shrinkage has been observed by different workers under steady state conditions even at temperatures well beyond 100°C. It is due to delayed diffusion, slow crystal growth and phase changes. Under transient temperature conditions it is difficult to separate thermal strains and deformations due to shrinkage. According to Bazant /6/ shrinkage can be estimated by

$$\varepsilon_s = k_s \cdot \Delta w \qquad (3.4)$$

whereby $k_s = 10^{-5} m^3/kg$ is independent of temperatures and Δw comprises the weight loss of drying cement paste. From Maréchal /7/ tests we have derived shrinkage data around 0.8 ‰ at 400°C. This value is being obtained if $\Delta w = 80$ kg/m³ according to Bazant's formula which seems to be a reasonable number if we assume average conditions for the concrete tested.

3.6 Transient State Creep, Restraint

Transient tests for measuring the total deformation or restraint of concrete have in principal the strongest relation to building fires and are supposed to give the most realistic data with direct relevance to fire. The tests yield strain-temperature relationships (parameter: load level α) for given heating rates. Numerous publications in this field of material research have appeared recently. Theoretical considerations and experimental investigations are reported in which the main features studied are the mix proportion, the type of aggregate, the curing condition and age of concrete and the test conditions (heating rate e.g.).

Fig. 16 shows the results of a total deformation test with a normal and a lightweight structural concrete. The test results are typical for concrete and the following conclusions can be derived:

- The water-cement ratio and the original strength are of little importance.
- The aggregate-cement ratio has a great influence on the shape of the strain-temperature curves and on the critical concrete temperature T_{cr}. The critical concrete temperatures of rich mixes are, at higher load levels, lower in comparison to lean mixes.
- The harder the aggregate and the lower its thermal expansion the lower the total deformations of normal concretes in transient tests. Lightweight concretes with expanded clay aggregates indicate the lowest total deformations.
- The curing conditions are of great importance in the temperature range of 20 to 300°C. Air cured and oven dried specimens indicate a significant lower transient creep than water cured specimens. The influence of moisture content on the total deformations at higher temperatures is negligible.
- The rate of heating is of minor influence as long as heating rates between 0.2 and 5.0 K/min are considered.

Restraining tests have been reported only in a few publications. The restraining and transient creep phenomena are complementary of each other. Therefore most of the statements regarding the transient creep data are (inversely) valid for the restraining data e.g. if a special type of concrete is supposed to have a low thermal expansion and a high transient creep it follows from theoretical considerations that relative low restraining forces are to be expected. Restraining forces are very sensitive to factors like moisture content, type of aggregate and the curing conditions etc. Typical restraining force-temperature curves are shown in fig.17.

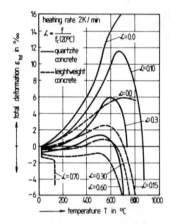 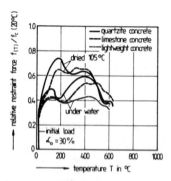

Fig.16: Total deformation on different concretes being loaded during heating up

Fig.17: Restraint forces of different concretes being totally restraint during heating up

The type of aggregate and the restraining forces suggest a close relationship. The behaviour of concretes made with different types of aggregates is in agreement with the results of the transient creep tests. At temperatures above 200°C sandstone and siliceous aggregate concretes attain the highest restraining forces due to their high thermal expansion. Lightweight or limestone aggregate concretes attain lower restraining forces due to a relatively lower thermal expansion.

The moisture content is a factor which determines the maximum value of the restraining forces in the temperature region between 20 to 200°C. With 100°C oven-dried specimens a maximum peak appears between approximately 100 to 200°C. The restaining forces attain values in the range of 60 to 80% of the ultimate strength at 20°C. The peak decreases rapidly. The behaviour of concrete with a high moisture content, i.e. after water storage, is quite different. The moisture favours higher creep deformations which result in a much lower restraining peak at 100°C compared with dried specimens. In the temperature range from 100 to 200°C a distinct minimum occurs in the restraining forces, which is connected with rapid drying and shrinkage effects. In this case the absolute maximum restraint appears at about 450°C. Beyond this temperature the restraining forces decrease due to increasing plasticity.The restraint of air-conditioned specimens (20°C/65% r.h.) is similar to that of the moist specimens.

4. ANALYTICAL MODELLING OF CONCRETE BEHAVIOUR

The modelling of concrete behaviour under high temperatures has been discussed during the recent 10 to 15 years. Since total deformation measurements were performed the debate concentrated on the question how the macroscopically measureable strains could be subdivided into individual strain elements. It is generally agreed that the total strain comprises three parts:

$$\varepsilon_{tot} = \varepsilon_M + \varepsilon_H + \varepsilon_T \tag{4.1}$$

in which ε_M = mechanical (elastic, plastic) strain, ε_H = hygral (shrinkage swelling) strain and ε_T = thermal dilatation /8/. The strains derived from total deformation tests do not distinguish between ε_H and ε_T i.e.

$$\varepsilon_{tot} = \varepsilon_M + \varepsilon_{th} \tag{4.2}$$

According to Khoury /9/ in this case the term ε_M is called "load induced thermal strain". It consists of transient creep (transitional thermal creep and drying creep), basic creep and elastic strains. Within the range of tested heating rates the load induced thermal strains turned out to be nearly independant of time, type of concrete, moisture and thermal expansion of concrete. Fig.18 shows the induced thermal strain curves of a one year old normal concrete and lightweight concrete. The strains indicate an approximately linear increase with an increase in the stress level. This observation is in agreement with older findings of Anderberg /10/ and Schneider /11/.

For stresses less than 0.5 of the strength limit Bazant /8/ proposed for creep at variable moisture content and temperatures:

$$\varepsilon_M = \sigma \cdot J (T, t, t') \tag{4.3}$$

in which $J (T,t,t')$ = compliance function, which represents the strain at age t caused by a unit stress that has been acting since age t' of concrete. The compliance function was approximated by the double power law

$$J (T,t,t') = \frac{1}{E_O} + g(w) \cdot \frac{f_w \cdot \phi_T}{E_O} \cdot f(t_e) \cdot (t-t')^{1/8} \tag{4.4}$$

in which E_O = elastic modulus, $g(w)$ = function for drying rates, f_w = function of water content, ϕ_T = function of temperature, $f(t_e)$ = function of maturity. The equation comprises all effects which are relevant for concrete behaviour under fire with two exceptions: The modulus of elasticity is not just temperature but also load dependent and ε_M turned out to be nearly time invariant at least within a time period of several hours i.e.

$$\varepsilon_M (T, t, t', \sigma) = \varepsilon_M (T, \sigma) \tag{4.5}$$

It is therefore convenient to write

$$\varepsilon_M = \varepsilon_{el} (T, \sigma) + \varepsilon_{pl} (T,\sigma) + \varepsilon_{tr,cr}(T,\sigma) \tag{4.6}$$

in which ε_{el} represents the elastic strains, ε_{pl} accounts for plastic strains due to stresses beyond 0.5 of the strength limit and $\varepsilon_{tr,cr}$ comprises all the other strain increments which occur under rapid heating and drying of loaded concrete and which is called transient creep /3/.

It should be noted that transient creep in this sense does not just comprise creep strains which are due to rapid drying of capillary water and connected with that an internal redistribution of moisture in the microstructure but it also accounts for the total loss of gel water and chemically bounded water. Both effects are usually called dehydration, i.e. transient creep holds for strains occuring during the change of matter until the strength or strain limits are exceeded. During the change of matter the microstructure converts into a solid with a considerable amount of internal micro and macro cracks.

In this connection it should be mentioned that Anderberg /10/ proposed a slightly different notation for the transient creep term:

$$\varepsilon_{tr,cr} = \varepsilon_{cr} + \varepsilon_{tr} \qquad (4.7)$$

in which ε_{cr} = creep strain measured under high (constant) temperatures, it comprises drying creep plus basic creep, and ε_{tr} = transient strain, accounting for the effect of the strain increase under increasing temperatures. As ε_{cr} is small with respect to fire situations the proposed formula yields no advantages in practical applications.

A compliance function which accounts for the three strain elements according to eq . (4.6) obtains the following form:

$$J (T, \sigma) = \frac{1}{E} (1 + \varkappa) + \frac{\Phi}{E} \qquad (4.8)$$

in which

$$E = E_o \cdot f(T) \cdot g(\sigma,T) \qquad (4.9)$$

whereby $f(T)$ may be a function according to fig. 9 and $g(\sigma,T)$ allows for the increase of elasticity due to external loads (comp. fig. 19):

$$g = 1.0 + \frac{\sigma(T)}{f_c(20°C)} \cdot (T-20) / 100 \qquad (4.10)$$

whereby the empirical boundary limit, if $\sigma/f_{c,20} > 0.3$ than $\sigma/f_{c,20} \equiv 0.3$ must be considered.

The $\varkappa$-function accounts for the plastic magnifications in a stress-strain diagram. It may be neglected within the elastic range i.e. load levels less than 0.5 of the strength limit. From the theoretical stress-strain relationship for concrete

$$\frac{\sigma(T)}{f_{ult}(T)} = \frac{\varepsilon(T)}{\varepsilon_{ult}(T)} \cdot \frac{n}{n-1 + (\frac{\varepsilon(T)}{\varepsilon_{ult}(T)})^n} \qquad (4.11)$$

in which $f_{ult}(T)$ = high temperature compressive strength according to fig.6 and $\varepsilon_{ult}(T) \equiv$ ultimate strains after fig.13 the following expression can be derived

$$\varkappa = \frac{1}{n-1} \cdot (\varepsilon(T)/\varepsilon_{ult}(T))^n \qquad (4.12)$$

whereby n = 2.5 for lightweight concrete and n = 3.0 for normal concrete. A reasonable approximation for $\varkappa$ is given by

$$\varkappa = \frac{1}{n-1} \cdot (\sigma(T)/f_{ult}(T))^5 \qquad (4.13)$$

A descending branch in the σ-ε diagram is not taken into account by the proposed eqs. (4.12) and (4.13). Creep functions φ have been derived by Schneider /11/ and are shown on fig. 19. The function Φ is described by

$$\Phi = g \cdot \phi + \frac{\sigma(T) \cdot (T-20)}{f_c(20°C) \cdot 100} \qquad (4.14)$$

whereby $\sigma(T)/f_c(20°C) \leq 0.3$ and

$$\phi = C_1 \cdot \tanh \, \gamma_w \, (T-20) + C_2 \tanh \, \gamma_o \, (T-T_g) + C_3 \qquad (4.15)$$

The function g is given in equ. (4.10) and γ_w accounts for the moisture content w in % by weight:

$$\gamma_w = (0.3 \cdot w + 2.2) \cdot 10^{-3} \qquad (4.16)$$

Fig. 19 is based on a moisture content of 2%. The overall influence is comparatively small. Table 4.1 contains the parameters for the φ functions of three structural concretes. The parameters are slightly modified compared to the results of older publications /3, 11/. Values above 800°C are derived by extrapolation.

Table 4.1: Parameters for transient creep functions φ of structural concretes

parameter	dimension	quartzite concrete	limestone concrete	lightweight concrete
C_1	1	2.60	2.60	2.60
C_2	1	1.40	2.40	3.00
C_3	1	1.40	2.40	3.00
γ_o	°C^{-1}	$7.5 \cdot 10^{-3}$	$7.5 \cdot 10^{-3}$	$7.5 \cdot 10^{-3}$
T_g	°C	700	650	600

It should be mentioned that nearly no transient creep does occur during cooling of concrete after a temperature exposure. A reasonable assumption for the concrete behaviour under cooling is that the prevalent modulus of elasticity is fixed according to the corresponding previous maximum temperature and stress states. Thereby it should be noted that thermal strains may be significantly irreversible under cooling conditions. This holds expecially for temperatures higher than 600°C and depends mainly on the type of aggregate in the concrete.

The failure of concrete may be determined by the maximum of attainable strains $\varepsilon_M \leq \varepsilon_{ult}$ which have been established in a total deformation test. Fig. 20 shows the results of structural concretes according to /3/. The strain limits are far beyond the values of common ultimate strains which were derived from stress-strain relationships (comp. fig.13).

68

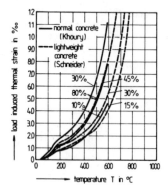

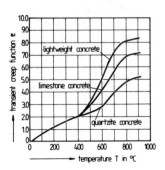

Fig.18: Load induced thermal strain
curves of normal and lightweight concrete

Fig.19: Transient creep function φ
for different concretes
derived from total deformation tests

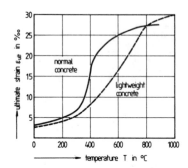

Fig.20: Ultimate strain of concrete
derived from the results of
total deformation tests

5. CONCLUSION

   Different models for determining the structural fire behaviour of concrete have been developed. The models use steady state, transient state and
mixed data and have different degrees of sophistication. As to be expected
transient state models correspond largely to fire situations.

According to recent research results concrete models should consider transient creep or at least appropriate strain effects. Especially if the theoretical calculation of concrete members requires the determination of de-

formations and restraint the consideration of transient effects is urgent. The proposed concrete model covers the most temperature effects which have been observed with concrete under fire and therefore can be used for practical design purposes.

6. REFERENCES

/ 1/ Bazant, Z.P.: Chern, J.C.

Normal and Refractory Concretes for LMFBR Applications. EPRI: NP-2437, Vol.1, Northwestern University, Evanston, 1982.

/ 2/ Kordina, K.:

The Behaviour of Structural Elements and Buildings under Fire (in German). Rheinisch-Westfälische Akademie der Wissenschaften, Nr. 281, Westdeutscher Verlag, Opladen, 1979.

/ 3/ Schneider, U.:

Properties of Materials at High Temperatures - Concrete -. RILEM 44-PHT, University of Kassel, Kassel, 1985.

/ 4/ Schneider, U.:

Behaviour of Concrete at High Temperatures. Deutscher Ausschuß für Stahlbeton, Heft 337, Verlag W. Ernst und Sohn, Berlin, 1982.

/ 5/ Ehm, C.:

Experimental Investigations of the Biaxial Strength and Deformation of Concrete at High Temperatures (in German). Dissertation, TU Braunschweig, 1985.

/ 6/ Bazant, Z.P.:

Mathematical Model for Creep and Thermal Shrinkage of Concrete at High Temperatures. Report No.82-10/ 249m, The Techn. Inst., Northwestern University, Evanston, 1982.

/ 7/ Maréchal, J.C.:

Le Fluage du Béton en Fonction de la Température. RILEM Colloquium. Mat. et Constr., Vol. 2, No.8, pp 111/115, 1969.

/ 8/ Bazant, Z.P.: Chern, J.C.

Concrete Creep at Variable Humidity: Constructive Law and Mechanism. Mat. et Constr., Vol.18, No.103, pp 1/20, 1985.

/ 9/ Khoury, G.A.

Strain of Concrete during first Heating Cycle to 600°C under Load. University of London, Submitted to Mag. of Concr. Research, May 1985.

/10/ Anderberg, Y.:

Fire-Exposed Hyperstatic Concrete Structures - An Experimental and Theoretical Study. Div. of Struct. Mechn. and Concrete Constr., Inst.of Techn., Lund, 1976.

/11/ Schneider, U.:

Creep and Relaxation of Concrete under High Temperatures (in German). Habilitation, TU Braunschweig, 1979.

STRENGTH AND FIRE ENDURANCE OF GLUED-LAMINATED TIMBER BEAMS

SCHAFFER, E.L., WOESTE, F.E., BENDER, D.A. AND MARX, C.M.

PFS Corporation, Madison, Wisconsin
    Virginia Polytechnic Institute
    Blacksburg, Virginia
            Texas A & M University
            College Station, Texas
                Alpine Engineered Products
                Pompano Beach, Florida

Research Sponsored By U.S. Forest Products Laboratory
Madison, Wisconsin
U.S.A.

ABSTRACT

The time-to-failure (TTF) of glued-laminated timber beams (glu-lam) is modeled by incorporating considerations for the mechanical properties of the laminating lumber, laminate length between finger joints, tensile strength of finger joints, and the thermal degradation of the beam during fire exposure (ASTM E-119). The model reasonably predicts both the room temperature strength and the TTF of loaded fire exposed glu-lam beams. The Monte Carlo computer simulation of the fabrication of many beams produced from randomly selected laminating stock, followed by strength or fire endurance analysis, additionally assesses the variability in performance. This is of import to fire safe reliability analyses.

INTRODUCTION

A number of techniques have been advanced world-wide to predict both the expected strength and fire endurance of timber beams. These methods have been developed for members of assumed uniform grade or having uniform mechanical properties in each laminate of glu-lam beams. As a result, the methods provide single valued estimates of beam strength or fire endurance. The growing trend to be able to evaluate the safety of structural members under service loads or fire exposure demands that variability in member response be described adequately. This paper discusses one such attempt to do so.

The discussion involves several key assumptions. The glu-lam beams are specifically loaded in bending and failure is initiated in the outer tension zone. Also, the glu-lam exhibits elastic behavior to failure.

Under fire exposure to ASTM E-119 conditions, the charring of the wood occurs at a constant rate, $\not\!\beta$, that is a function of species, specific gravity and moisture content of the wood.

METHODOLOGY 1|
Required inputs to the procedure are grade order, species of the laminations and lumber lengths in the beam, distributions of MOE and tensile strength, and distributions of end-joint tensile strength. For each of these beams the gross MOE, ultimate moment, apparent MOR, failure location, and failure mode are calculated. The following steps provide a summary of the simulation procedure for a single beam. These steps are repeated according to the desired sample size.

1. Laminating lumber of random length, MOE, and tensile strength are dealt into the bottom layer of the beam until the beam length requirements have been satisfied. Each subsequent layer is generated according to the layup of the beam by grade and species.

(Corresponding tensile strength values for the MOE-tensile strength [E-T] pairs were calculated based on regression equations determined from earlier data and described by Bender et al [1984].)

2. A random end-joint tensile strength is assigned wherever two pieces of lumber meet end-to-end. End joints in the simulated assembly are not allowed to occur within 6 inches of each other on adjacent tension lamination in conformance with Voluntary Product Standard PS-56 (1973).

3. Steps 1 and 2 are repeated until the entire beam has been assembled. All of the random strengths and stiffnesses are recorded in arrays, along with the location of each end joint.

4. A transformed section analysis (Brown and Suddarth 1977) is repeated across the entire beam at a specified increment of beam length. In each case, tensile strength and stiffness are used to calculate the ultimate moment. Then, the ultimate moment, gross MOE, failure location, and failure mode are stored in an array.

5. A transformed section analysis is performed at each end-joint location. End-joint tensile strength and stiffness are used to calculate the ultimate moment. The end-joint stiffness is taken to be the average stiffness of the two connecting pieces of lumber.

6. The minimum value of the ultimate moments of steps 4 and 5 is recorded. This value defines the ultimate moment-carrying capacity for the assembled glu-lam beam. These moments are compensated for the location on the moment diagram for two-point loading.

1| The Computer Program, plus example cases, is available in the Bender et al (1984) publication.

7. The apparent MOR is calculated by assuming a homogeneous beam cross section.

8. The ultimate moment-carrying capacity of the beam and the associated MOR, gross MOE, failure location, and failure mode are recorded.

A summary of the input parameters used in the predictive model is given in table 1. The simulated fabrication and corresponding bending test was normally conducted with 500 randomly simulated beams per beam geometry and load.

To fully simulate the behavior of actually tested beams, a correction to the apparent tensile strength, T, of the laminates must be made to compensate for the actual length of beam under constant moment. For example, the span between supports may be 25.5 feet and between the two load points 5 feet for two-point loaded beams. The 5-foot separation between the load points is essential to transforming a known tensile strength, T, of a long piece of lumber to one, T', shorter in length. This transformation was given previously by Bender et al (1984).

$$T' = T(N)^{1/\eta} + \mathcal{A}[1 - (N)^{1/\eta}] \qquad (1)$$

The parameters η and $\mathcal{A}$ are the shape and location parameters for the three-parameter Weibull distribution of tensile strength. The value N equal to 2.4 is computed by dividing the distance between grips of the experimentally tested tensile specimens used to develop E-T regression parameters (12 ft.) by the distance between the two points of load application of the test beams (5 ft.). Further discussion of how this transformation was derived is given in a previous report (Bender et al 1984).

STRENGTH SIMULATIONS AND EXPERIMENTAL RESULTS

Experimental Beams -- Twenty-one (21) Douglas-fir test beams were manufactured by a commercial laminator following their normal plant procedures. The beams were certified as conforming to the Voluntary Product Standard PS-56-73 for Structural Glued Laminated Timber. A horizontal finger joint (fingers visible on the narrow face) was used and bonded with a melamine adhesive. A phenol-resorcinol adhesive was used for face gluing.

After gluing, the beams were segregated into three groups of seven each; groups A, B, and C. The group A beams were planed to 5.12 by 16.5 inches, the group B beams to 3.62 by 15.75 inches, and the group C beams to 2.12 by 15.0 inches to simulate residual cross-sections before and after 30 minutes and 60 minutes respectively of ASTM E 119 3-sided fire exposure (figure 1). For all groups, equal amounts were planed from both sides of the beam. Half of the tension lamination thickness in group B beams was removed by a sanding operation.

TABLE 1: Coast Douglas-fir input parameters for the beam strength prediction model include Weibull parameters for modulus of elasticity data, regression parameters which relate tensile strength to modulus of elasticity, and log-normal parameters used to define the length of the pieces of lumber. Modulus of elasticity Weibull scale and location parameters (∇ and ψ) must be multiplied by 1 million. Also included are Weibull parameters for the tensile strength of horizontal finger joints ("L" and "S" sets). Tensile strength Weibull scale and location parameters (∇ and ψ) must be multiplied by 1,000.

Lumber Grades	Sample size	Weibull parameters[1] Scale (∇) Lb/in.^{2}	Shape (η)	Location (ψ) Lb/in^2	Regression parameters[2] B_0	B_1	K	Length parameter[3] λ	ξ
		LAMININATE MODULUS OF ELASTICITY							
302-24	40	0.89	2.44	1.82	7.71	0.306E-06	0.325E-07	2.283	0.198
L1	55	.72	2.47	1.69	6.83	.561E-06	.298E-07	2.283	.198
L2D	79	1.05	3.71	1.18	6.58	.678E-06	.231E-07	2.471	.166
L2	74	1.03	4.54	.99	6.58	.678E-06	.231E-07	2.471	.166
L3	156	.97	3.91	.78	6.88	.470E-06	.336E-07	2.471	.166
		FINGER JOINT TENSILE STRENGTH							
"L" set		3.81	3.55	3.35	--	--	--	--	--
"S" set		2.18	2.09	3.01	--	--	--	--	--

1 The Weibull parameters were determined based on the data collected on the lumber used to build experimental beams.
2 The regression parameters were determined from earlier E-T paired data and described by Bender et al (1984).
3 Assumed shorter lumber lengths on the average for the higher quality grades.

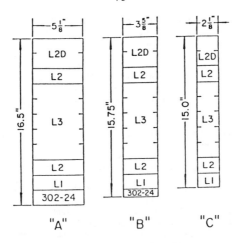

Figure 1--Douglas-Fir Larch test beams. There were seven beams in each of these test groups (M151 353).

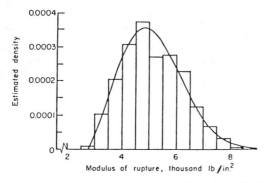

Figure 2 - Typical histogram and estimated three-parameter Weibull density functions of modulus of rupture for simulated type A Douglas-fir beam (5-1/8-by 16-inch cross section). Finger-joint distributions used as input were for the "S" set.

Simulation Comparisons -- Two sets of input finger-joint strength data were used. The first set was termed "L" to represent end-joint derived strengths for a large population of industry-wide generated Douglas-fir end joints. Similarly, the second set is termed "S" to represent a smaller set of finger joints collected at the same time the experimental beams were fabricated. The strength results of beam simulations as portrayed by the three-parameter Weibull distribution parameters (table 2) and resulting mean and COV are shown in table 3 for the three beam types shown in figure 1.

Figure 2 shows a typical histogram of the predicted MOR for the beam A group simulated having end joints of the "S" joint set. Superimposed over the data is curve corresponding to the three-parameter Weibull density function of the predicted MOR.

Illustrated in figure 3 are the mean MOR's for the simulated and experimentally observed three beam types. Note that the observed values are greater than that predicted using either the "L" or "S" group joint strengths for beam types B and C. For type A, the mean strength falls between set "L" and "S" predictions, but is closer to set "L". It can also be observed that the experimental results fall well within the 90 percent confidence intervals for the "L" set predicted means.

It is well known that both lamination and finger-joint quality are the key factors controlling glued laminated beam strength in practice. Our results are in agreement with that observation. It is clear from the correlation of beam experimental strength results to those predicted that predictive capability is sensitive to both the lamination strength and strength of the finger joint.

FIRE ENDURANCE SIMULATIONS

A fire endurance prediction model was developed by making refinements to the beam strength model. The fire endurance of glu-lam beams is measured by the time-to-failure (TTF), where TTF is defined as the length of time that a structure will support its design load when subjected to intense fire conditions. The fire endurance model can be used to predict the distribution of TTF for any glu-lam beam of interest. Fire is simulated by removing the char layer from the beam cross section. The thickness of the char layer, R, is given by

$$R = \beta t + \delta \qquad (2)$$

Where

β = char rate

t = fire exposure time

δ = finite thickness of residual wood which is weakened by the elevated temperature and moisture.

TABLE 2: Parameters for three-parameter Weibull distribution of predicted
beam strengths. Two different assumptions of finger-joint strength
distributions are employed in the predictive model--set "L" and "S".

Beam Type	Set "L"[1]			Set "S"[1]		
	Scale (σ)	Shape (η)	Location (μ)	Scale (σ)	Shape (η)	Location (μ)
	$Lb/in.^2$		$Lb/in.^2$	Lb/in^2		$Lb/in.^2$
A	3,020	3.53	3,590	2,650	2.28	2,770
B	2,660	3.78	3,120	2,460	2.58	2,390
C	2,150	3.29	2,960	2,240	3.17	2,660

[1] "L" set: Employing finger-joint strength data from estimated whole
population properties. "S" set: Employing finger-joint strength data
estimated for beams fabricated.

TABLE 3: Means and coefficients of variation (COV) for experimental and
predicted modulus of rupture strengths for three glu-lam beam sets.

| Beam Type | Experimental | | Predicted[1] | | | |
			"L" Set		"S" set	
	Mean [2]	COV	Mean	COV	Mean	COV
	$Lb/in.^2$	Pct	$Lb/in.^2$	Pct	$Lb/in.^2$	Pct
A	5,980	22.0	6,230	13.5	5,120	21.4
B	6,080	27.9	5,520	12.7	4,580	20.0
C	5,630	13.9	4,890	13.0	4,670	14.8

[1] "L" set: Employing finger-joint strength data from estimated whole
population properties. "S" set: Employing finger-joint strength data
estimated for beams fabricated.

[2] unadjusted means.

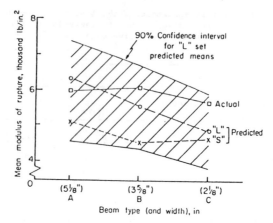

Figure 3 – Predicted and actual mean strengths
as a function of beam type (and width of beam).

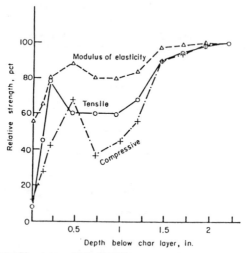

Figure 4 – Relative modulus of elasticity, compressive
and tensile strength as a function of distance below
char layer in softwood section under fire exposure.
(Expressed in percent of that at 25° C and initial
moisture content of 12 percent.) Duration of
fire exposure should be equal to or greater than
20 minutes to apply results of this figure (M149–821)

Two of the assumptions of this model are that char rate, β, and residual thickness, δ, remain constant. These two assumptions have received considerable experimental support (Schaffer 1960 & 1965). Another assumption is that the unit strength and E properties of individual laminations remain constant as the cross section is reduced.

The following steps summarize the simulation procedure of the fire endurance model for a single beam. These steps are repeated for each beam in the simulation:

1. A beam is randomly fabricated in a computer simulation in the same manner as that for strength assessment.

2. Full design load (that develops the full allowable stress in bending) is applied. Allowable stress is compensated for beam size.

3. A transformed section analysis is performed along the entire length of the beam at specified increments of length and at each finger-joint location.

4. Computed stress levels are compared with corresponding tensile strength values. If tensile strength is exceeded, failure occurs and TTF is recorded.

5. Critical moment permitted by lateral torsional buckling is calculated employing the lateral buckling equation derived by Pettersson (1952). If the critical moment is exceeded by the applied moment, failure occurs and the TTF noted.

6. Time is increased 1 minute and the corresponding char thickness as produced by a standard fire exposure per American Society for Testing and Materials Standard ASTM E 119 (1979) is removed from the cross section.

7. Steps 3 through 6 are repeated until beam failure occurs.

8. Steps 1 through 7 are repeated for each beam simulated.

We previously identified the parameters critical to predicting room temperature beam strength. There is, however, a need to explain the origin of parameters necessary to the analysis of strength under fire exposure. There are:

1. Conversion of the beam span, L_s, under two-point concentrated load to an effective span L_s^*, under uniform load common in fire exposure.

2. The depth of char as a function of char rate, β, and time, t.

3. Compensation for the loss of strength of the heated uncharred wood by assuming an additional layer of wood, δ, has zero strength.

Effective span, L_s*, --The predictive model for beam strength and
fire endurance is sensitive to whether the load is applied uniformly
or concentrated at several points. To consistently compensate for
this difference one should convert the span under concentrated load,
L_s, to an effective length, L_s*, under uniform load. From beams
tested under two-point loading a span to beam depth (d) ratio of at
least 15:1 is recommended to minimize the likelihood of failure in
horizontal shear. As a result, the effective span length, L_s*, for
a uniformly loaded beam was defined as:

$$L_s* = L_s - 15 \, d \qquad\qquad (3)$$

This effective span is further employed in a transformation
equation (2) to determine the expected tensile strength of the
laminating lumber between finger joints as compared to that determined
for experimentally tested lumber of fixed length.

Char rate, β, --The char rate is assumed to be an average 0.025
inch/minute for Douglas-fir. Though this char rate could be entered
as a random variable as well, it was a constant in this investigation.

Zero-strength layer, $\beta t + \delta$.--The char layer forms at a steady
rate, β, under standard fire exposure. Because it is highly porous
and fissured, it can be assumed to have nonload-carrying ability. The
wood below the char layer is heated, however, and moisture moves
through it as it dries. Both temperature and moisture content affect
wood strength and stiffness. For Douglas-fir, in large sections such
as glued laminated beams and columns, the temperature achieves a
quasi-steady-state distribution below the charwood interface (Schaffer
1977). The moisture gradient has also been experimentally
investigated for Douglas-fir at 12 percent moisture content by White
and Schaffer (1981).

To compensate wood strength and MOE for temperature and moisture
content, procedures are available (Schaffer 1984). If the tensile,
compressive, and MOE values are computed for uncharred wood in a
fire-exposed section, the response is as shown in figure 4 (Schaffer
1984). These properties are most affected within a 1.5 inch (40 mm)
layer below the char-wood interface in a fire-exposed section.

To simplify analysis of beam fire endurance, it was attractive to
examine whether a layer of uncharred wood of thickness, δ, for which
negligible strength is assumed, might be subtracted from the beam
cross-sectional dimensions in addition to the char layer thickness,
βt. This was done by (a) averaging the tensile, compressive, and
MOE response over the 1.5 inch (40 mm) heated layer, (b) analyzing the
beam using transformed section analysis, and (c) ascribing the full
loss of strength to a heated layer of thickness, δ.

The mean strength properties and variation for the 1.5-inch (40-mm) heated layer are expressed as a percent of that at room temperature and wood moisture content of 12 percent as follows:

	Mean	COV
Tensile strength	66.1 pct	14.5 pct
Compressive strength	54.4 pct	19.3 pct
MOE	83.4 pct	9.0 pct

Analyzing a three-sided simulated fire-exposed beam for strength using transformed section analysis resulted in the observation that the effective mean tensile strength is reduced to a level 79.3 percent of that at room temperature. Using this information, it was estimated that a layer δ , 0.3 inch thick, can be deducted from the heated zone beneath the char on fire-exposed sides of the beam and the residual strength with duration of fire exposure estimated (figure 5).

COMPARISON OF FIRE ENDURANCE SIMULATIONS

Employing the "L" set parameters for values of finger-joint strength and typical properties of lamination grades of Douglas-fir, the fire endurance (TTF) was analyzed for 5.125- by 16.5-inch 11-lamination Douglas fir-Larch beam (24F-V4) of 25.5-foot span and carrying full allowable uniform load (47.7 lb/in.). Three-sided fire exposure was assumed; however, four-sided exposure can also be accommodated.

The simulated random fabrication and analysis of the TTF under fire exposure for 100 beams was performed. The mean TTF was estimated as 35.2 minutes with a COV of 13.7 percent. Lateral torsional buckling was never the cause of failure in any of the repeated simulations. If the bottom lamination is defined as the first lamination, simulated failure normally initiated in the second lamination (57/100), but was found to occur in the first (25/100) and third (18/100) as well. The charring of about half the thickness of the first lamination contributed to, and explains much of, this result.

The TTF for three-sided fire exposure of beams of the same dimension and subjected to full design load was also calculated using two other methods by Lie (1977) and Meyer-Ottens (1979). Both assume that a model applies to the form

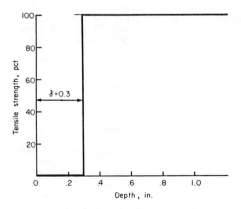

Figure 5 – Simplified diagram attributing all tensile
strength loss in a charring beam to a thickness of 0.3 inch
into the heated wood below the advancing char.

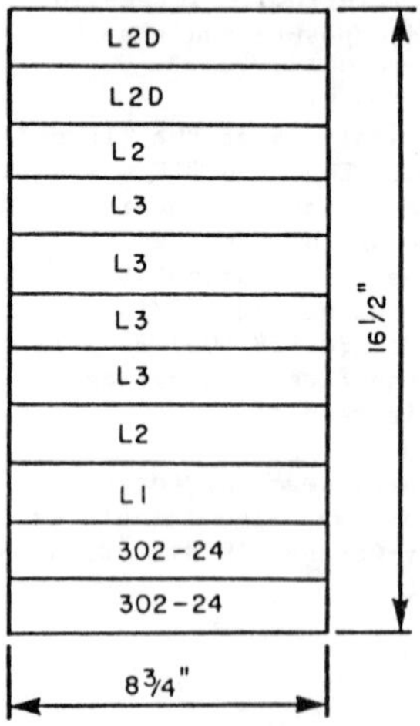

L 2 D
L 2 D
L 2
L 3
L 3
L 3
L 3
L 2
L I
302 – 24
302 – 24

$16\frac{1}{2}"$

$8\frac{3}{4}"$

Figure 6 – Layup of Douglas Fir -- Larch glu-lam
beam experimentally fire endurance tested.

$$M/S(t) = \alpha \; \sigma_{cr}$$

where   M   = the maximum applied moment

          S(t) = the section modulus of the uncharred wood section
                  as a function of duration, t, of fire exposure.

          σ_{cr} = strength of wood at room temperature.

          α = the ratio of strength of the outer uncharred wood
                  fibers of the beam to that at room temperature.

The factor, α , varies from 0.5 to 0.8 in international literature.

     The TTF result obtained using Lie (1977) is predicted at 33.2
minutes and that employing Meyer-Ottens (1979) about 30 minutes. A
more refined analysis provided by A. Haksever (1980) that expands upon
Meyer-Ottens (1979) indicated the TTF is 31.2 minutes.

     The above results are within the 65 percent confidence band
(30.4-40.0 min) of the model for fire endurance presented in this
paper. Hence the results are similar. Only the results, however, of
Meyer-Ottens (1979) and Haksever (1980) have a strong experimental
fire endurance data base. A total of 35 glu-lam beams were tested
under load and fire exposure. Unfortunately, the design stress
statistics are not reliably known so that the applied load can be
compared directly with those allowed in North American practice. It
appears, however, that West Germany applies a smaller general
adjustment factor on the 5 percent exclusion limit for MOE (Goodman
1980). In North America the factor is 2.1 and in Germany, evidently,
is 1.88. In essence, then, 11.7 percent higher allowable design
stresses exist in Germany as compared to North America. This is
translated to 11.7 percent higher design load being applied in German
fire endurance tests of timber beams as compared to North America. An
1.7 percent decrease in developed beam stress can increase the TTF of
a 5.12- by 16.5-inch glu-lam beam (S = 3,810 cm^3) about 7 minutes
(Meyer-Ottens 1979). This results in a North American compatible fire
endurance estimate of 38.2 minutes.

     The predicted TTF of 35.2 minutes developed by the model in this
paper is seen to fall between the 33.2 minutes and 38.2 minutes as
estimated from these sources and observed to be well within the
confidence limit of $\pm$ 4.8 minutes obtained by the simulation model.

     The United States and Canada recently conducted single glu-lam
beam fire endurance test. The 11-lamination Douglas fir-Larch beam
(24F-V4, 10-lmaination design plus additional 302-24 grade tension
lamination) was 8.69 by 16.44 inches in cross section (figure 6). The
beam span was 16.97 feet center to center of bearing. It was loaded
to 71.5 percent of full design load. The adjusted full allowable
bending stress for this section was 2,396.5 lb/in^2 and the load was
applied at the three quarter points.

Prior to testing, the TTF was simulated, using the model developed here, for 100 randomly generated beams. Fifty were subjected to full design load and 50 to 71.5 percent of full design load. The calculated mean TTF and standard deviation were:

	Full load (min)	71.5 percent load (min)
Mean TTF	70.5	86.0
Standard Deviation	10.2	7.5

The beam was exposed to ASTM E 119 (1979) fire exposure on three sides. Failure to carry the load (rupture) occurred at 86.25 minutes.

The comparison of predicted to observed results shows that the experimental result fortuitously coincides with the mean TTF for those simulated, and that a typical beam fully loaded could expect to have a TTF of 70.5 minutes. In addition, 95 percent of the beams so tested under full load would be expected to have a TTF greater than 53.7 minutes!

CONCLUSIONS

A reliability-based computer model employing a Monte Carlo simulation technique as previously developed, is used to predict the variation in strength and fire endurance of glu-lam beams (Bender et al 1984). One may conclude that:

1. The predictive model is a slightly conservative predictor (underestimator) of beam strength as based upon the rational analysis of independent test results.

2. An 11-lamination Douglas-fir glu-lam beam of 5.125- by 16.5-inch cross-section and of 24F-V4 design is predicted to have a time-to-failure under standard fire exposure (ASTM E-119) of 35 minutes (COV of 13.7 pct) while carrying full design load. Predicted results compare well with world literature.

3. The model accurately predicted the results of one well-controlled full-size glu-lam beam fire dndurance test. The time-to-failure under 71.5 percent of full design load was predicted to have a mean of 86.0 minutes (COV of 7.5 pct) and the actual beam failed at 86.25 minutes. (Under full design load, the model estimated mean time-to-failure as 70.5 minutes.)

4. The simulation model, as in the bending strength case, provides predicted mean TTF estimates for beam types as well as the COV. Information of this type can be used in a second-moment reliability analysis for fire safety if variation in the fire exposure severity and applied load are additionally available.

LITERATURE CITED

American Institute of Timber Construction. Manufacturing specifications for structural glued laminated timber of softwood species. Design and Manufacturing. AITC 117-82. Englewood, CO: AITC; 1982.

American Institute of Timber Construction. Voluntary Product Standard PS-56-73 for structural glued laminated timber. U.S. Department of Commerce; 1973.

American Society for Testing and Materials. Standard methods of static test of timbers in structural sizes. ASTM Standard D 198-76. Philadelphia, PA: ASTM; 1976.

Bender, D.A.; Woeste, F.E.; Schaffer, E.L.; Marx, C.M. Reliability formulation for strength and fire endurance of glued-laminated beams. (USDA Forest Service, Forest Products Laboratory Research Paper FPL 460) 1984.

Brown, K.M.; Suddarth, S.K. A glue laminated beam analyzer for conventionasl reliability based engineering design (RB 94). West Lafayette, IN: Department of Forestry and Natural Resources, Wood Research Laboratory; 1977.

Goodman, James. Private communication (7/21/80). Fort Collins, CO: Colorado State University, Department of Civil Engineering; 1980.

Haksever, A. Private communication (12/80). Institute fur Baustoffe, Massivbau and Brandschutz. Braunschweig, West Germany: Technische Universitat Braunschwig; 1980.

Lie, T.T. A method for assessing the fire resistance of laminated timber beams and columns. Canadian Journal of Civil Engineers. 4(2): 161-169; 1977.

Meyer-Ottens, C. Feuerwidenstands dauer unbekleiderter hoher REchteck balken aus Brett schichtholz. Forschungsbeitrage fur die Baupraxis. Berlin: Wilhelm Ernst and Sohn; 1979.

Pettersson, O. Combined bending and torsion of I beams of monosymmetrical cross-section. Bull. 10. Stockholm, Sweden; Divison of Building Statics and Structural Engineering, Royal Institute of Technology; 1952.

Schaffer, E.L. Approach to mathematical prediction of temperature rise within a semi-infinite wood slab subjected to high-temperature conditions. Pyrodynamics. 2: 117-132; 1965.

Schaffer, E.L.   State of structural timber fire endurance.   Wood and Fiber.   9(2):   145-170; 1977.

Schaffer, E.L.   Structural fire design: Wood. Res. Pap. FPL 450. USDA Forest Service, Forest Products Laboratory, Madison, WI;

Schaffer, E.L.   Charring rate of selected woods transverse to grain. Research Paper FPL 69, USDA Forest Service, Forest Products Laboratory, Madison; WI, 1967.

White, R.H.; Schaffer, E.L.   Transient moisture gradient in fire-exposed wood slab.   Wood and Fiber.   13(1):   17-38; 1981.

PROPERTIES OF BUILDING MATERIALS: BASES FOR FIRE SAFETY DESIGN

HARMATHY, TIBOR Z.

Fire Research Section,
Division of Building Research,
National Research Council Canada,
Ottawa, Ontario, Canada, K1A OR6

ABSTRACT

Some properties of building materials are structure-sensitive, some others structure-insensitive. Degree of crystallinity, grain structure, and pore structure are the most important microstructural characteristics that affect material behaviour. Specific (internal) surface and porosity are the factors that determine the moisture sorption characteristics of porous materials.

From the point of view of the role building materials play in the structural performance of buildings in fire, they can be classified as load-bearing (L), load bearing/insulating (L/I), or insulating (I). The designer is interested in the mechanical properties of materials in the L group, in both the mechanical and thermal properties of those in the L/I group, and the thermal properties of those in the I group. Of particular interest are: among mechanical properties – stress-strain curve, modulus of elasticity, yield strength, ultimate strength, and creep; among thermal properties – dilatometric curve, thermogravimetric curve, density, porosity, calorimetric curve, and thermal conductivity. Available information on these properties of building materials in all three groups is reviewed.

INTRODUCTION

Thanks to rapid advances in several fields of fire science, reliance on performance tests in fire safety design has diminished steadily during the past twenty years. Design for fire resistant compartmentation is one of the fields where noteworthy progress has been recorded. Engineers and architects, taking advantage of greatly improved computational techniques and facilities, are now able to make decisions with respect to measures to be taken to reduce the potential of fires to spread by destruction. The calculation of the fire resistance of many types of building elements has become a reality. Yet, there is still some reluctance by the writers of building codes to authorize the wholesale use of calculation results. Their reluctance has probably stemmed from an underlying doubt about the accuracy of input information into those calculations: the properties of building materials at elevated temperatures. Since techniques for the calculation of fire resistance are available at all levels of sophistication, it indeed appears that real progress in fire science will be measured for many years to come in terms of progress in materials science.

This paper will give an outline of how much is already known about building materials, and point out some areas where information is still hard to come by.

THE NATURE OF SOLID MATERIALS

Homogeneous materials, i.e., materials that have the same composition and the same properties in all their volume elements, are rarely found in nature. The heterogeneity of some construction materials, e.g., concrete, is easily noticeable. Other heterogeneities, those related to the microstructure of materials, i.e., their grain and pore structures, are not detectable by the naked eye. The microstructure depends greatly on the way the materials are formed. In general, those formed by solidification from a melt show the highest degree of homogeneity. The result of the solidification is normally a polycrystalline material, which comprises polygonal grains of crystals of random orientations. Severe cold-working in metals may produce an elongated grain structure and crystals with preferred orientations.

If cooled rapidly, some non-metallic materials may assume an amorphous or glassy structure. Glassy materials have a liquid-like, grainless microstructure with low crystalline order. On heating they will go through a series of phases of decreasing viscosity.

Synthetic polymers are made up of very large molecules. In the case of thermoplastics, on heating the molecular chains become more mobile relative to one another. The material softens like glasses do. In thermoset materials cross-bonds between the molecular chains prevent the loosening of the molecular structure and their change into a liquid-like state.

Some building materials are formed from a wet, plastic mass or from compacted powders by firing. The product is a polycrystalline solid with a well-developed pore structure.

Two important building materials, concrete and gypsum, are formed by mixing finely ground powders (and aggregates) with water. The mixture solidifies by hydration. The cement paste in a concrete has a highly complex microstructure interspersed with very fine, elaborate pores.

In dealing with practical fire safety problems, most building materials can be treated as isotropic materials, i.e., as though they possessed the same properties in all directions. Among their properties, those that are well-defined by the composition and phase are structure-insensitive. Some others depend on the microstructure of the solid or on its previous history. These properties are structure-sensitive.

Unlike some metals and ceramic materials, building materials are not usually required to function satisfactorily at elevated temperatures for an extended period. On heating, most of them undergo physicochemical changes, accompanied by transformations in their microstructure and, at the same time, changes in their properties. A concrete at 500°C is completely different from what it is at room temperature. Clearly, the generic

information available on the properties of building materials at room
temperature is seldom applicable in the design for fire safety.

POROSITY AND MOISTURE SORPTION

If the material is porous — and most building materials indeed are — it
consists of at least two phases: a solid-phase matrix, and a gaseous-phase,
namely air, in the pores within the matrix. Usually, however, there is a
liquid or liquid-like phase also present: moisture either adsorbed from the
atmosphere to the pore surfaces, or held in the pores by capillary
condensation. This third phase is always present if the pore structure is
continuous. Discontinuous pores (like the pores of some foamed plastics)
are not readily accessible to atmospheric moisture.

The pore structure of materials is characterized by two properties:
porosity, P (m^3/m^3), the volume fraction of pores within the visible
boundaries of the solid; and specific surface, S (m^2/m^3), the surface area
of the pores per unit volume of the material. For a solid that has a
continuous pore structure, the porosity is a measure of the maximum amount
of water the solid can hold when saturated. The specific surface and (to a
lesser degree) porosity together determine the moisture content the solid
can hold in equilibrium with given atmospheric conditions (1).

The sorption isotherm shows the relationship at constant temperature
between the equilibrium moisture content of a porous material and the
relative humidity of the atmosphere. A sorption isotherm usually has two
branches: an adsorption branch obtained by monotonically increasing the
relative humidity of the atmosphere from 0 to 100% through very small
equilibrium steps, and the desorption branch, obtained by monotonically
lowering the relative humidity from 100 to 0%. Derived experimentally, the
sorption isotherms offer some insight into the nature of the material's pore
structure (1).

Among the common building materials, only two, concrete (more exactly,
the cement paste in the concrete) and wood, on account of their large
specific surfaces, can hold water in substantial enough amounts to be taken
into consideration in fire performance assessments.

REFERENCE CONDITION

Since the presence of moisture may have a significant and often
unpredictable effect on the properties of materials at any temperature below
100°C, it is imperative to conduct all property tests on specimens brought
into a moistureless "reference condition" by some drying technique prior to
the test. The reference condition is normally interpreted as that attained
by heating the test specimen in an oven at 105°C until its weight shows no
change. A few building materials, however, among them all gypsum products,
may undergo irreversible physicochemical changes when held at that
temperature for an extended period. To bring them to a reference condition,
specimens of these materials should be heated in a vacuum oven at some lower
temperature level (e.g., at 40°C in the case of gypsum products).

MIXTURE RULES

Some properties of materials of mixed composition or mixed phase can be calculated by simple rules if the material properties for the constituents are known. The simplest mixture rule is (2):

$$\pi^\zeta = \sum_i v_i \pi_i^\zeta \tag{1}$$

where π is a material property for the composite, π_i is that for the i-th constituent of it, v_i (m^3/m^3) is the volume fraction of the i-th constituent in the composite, and ζ has a value between -1 and $+1$.

Hamilton and Cosser (3) recommended the following rather versatile formula for two-phase solids:

$$\pi = \frac{v_1 \pi_1 + \alpha v_2 \pi_2}{v_1 + \alpha v_2} \tag{2}$$

where

$$\alpha = \frac{n \pi_1}{(n - 1)\pi_1 + \pi_2} \tag{3}$$

Here, phase 1 must always be the principal continuous phase, n is a function of the phase distribution geometry and, in general, has to be determined experimentally. With $n \to \infty$ and $n = 1$, eq. (2) converts to eq. (1) with $\zeta = 1$ and $\zeta = -1$, respectively. With $n = 3$, a relation is obtained for a two-phase system where the discontinuous phase consists of spherical inclusions (4).

By repeated application, eqs. (2) and (3) can be extended to a three-phase system (5), e.g., to a moist, porous solid, which consists of three essentially continuous phases (the solid matrix, with moisture and air in its pores).

CLASSIFICATION OF BUILDING MATERIALS

There are combustible and noncombustible building materials. Combustible building materials may become ignited in fire and burn. Information on those properties of combustible building materials that are related to their ignition and burning is woefully inadequate and therefore will remain outside the scope of this paper.

From the point of view of the role materials play in the structural performance of buildings in fire, the following grouping appears to be appropriate:

(1) Group L (load-bearing materials): materials designed to carry high stresses, usually in tension. Clearly, for group L materials the mechanical properties related to their behaviour in tension are of principal interest to the designer.

(2) Group L/I (load-bearing/insulating materials): materials designed to carry moderate stresses and, in fire, to provide insulation to group L materials. For group L/I materials both the mechanical properties (related mainly to behaviour in compression) and the thermal properties are of interest.

(3) Group I (insulating materials): materials not designed to carry load. Their role in fire is to resist heat transmission through building elements and/or to provide insulation to group L materials. For group I materials only the thermal properties are of interest.

Among thousands of building materials in use, the following deserve special attention:

- in group L: structural steel and prestressing steel,
- in group L/I: concrete, brick, and wood,
- in group I: gypsum.

MECHANICAL PROPERTIES

Stress-Strain Curve

The mechanical properties of solids are usually derived from conventional tensile or compressive tests. The result of such a test is a stress-strain curve (Figure 1), showing the variation of stress, σ (MPa), with increasing deformation (or strain), ε (m/m), while the material is deformed (strained) at a more or less constant rate (i.e., constant crosshead speed), usually of the order of 1 mm/min.

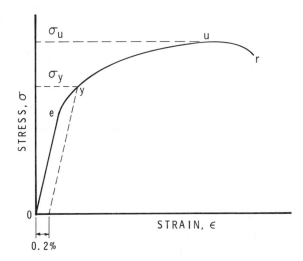

Figure 1. Stress-strain curve (strain rate is roughly constant)

Modulus of Elasticity, Yield Strength, Ultimate Strength

Section 0-e of the curve in Figure 1 represents the elastic deformation of the material, which is instantaneous and reversible. The modulus of elasticity, E (MPa), is the slope of that section. Between points e and u the deformation is plastic, nonrecoverable, and quasi-instantaneous. The plastic behaviour of the material is characterized by the yield strength at 0.2% offset, σ_y (MPa), and the ultimate strength, σ_u (MPa). After some localized necking (reduction of cross-sectional area), the material ruptures at point r.

The modulus of elasticity is more or less a structure-insensitive property.

For steels of similar metallurgical characteristics, the stress-strain curve does not depend significantly on the crosshead speed. At elevated temperatures the material undergoes plastic deformation even at constant stress, and the e-r section of the stress-strain curve depends markedly on the crosshead speed.

Creep

The time-dependent plastic deformation of the material is referred to as creep strain, and is denoted by ε_t (m/m). In a creep test the variation of ε_t is recorded against time, t (h), at constant stress (more accurately, at constant load) and at constant temperature, T (K). A typical strain-time curve is shown in Figure 2a. The total strain is

$$\varepsilon = \frac{\sigma}{E} + \varepsilon_t \quad (\sigma \approx const) \quad (4)$$

The 0-e section of the stress-time curve represents the instantaneous elastic (and recoverable) part of the curve; the rest is creep which is essentially nonrecoverable. The creep is fast at first (primary creep, section e-s_1 in the figure), then it proceeds for a long time at an approximately constant rate (secondary creep, section s_1-s_2), and finally it accelerates until rupture occurs (tertiary creep, section s_2-r). The curve becomes steeper if the test is conducted either at a higher load (stress) or at a higher temperature.

Dorn's creep concept (6) is particularly suitable for dealing with deformation processes developing at varying temperatures. Dorn eliminated the temperature as a separate variable by combining it with time,

$$\theta = \int_0^t e^{-\Delta H/RT} dt \quad (5)$$

where θ is temperature-compensated time (h), ΔH is the activation energy of creep (J/(kmole)), and R is the gas constant (J/(kmole)K).

From a practical point of view, only the primary and secondary creeps are of importance. The author showed (7, 8) that the creep strain in these two regimes can be satisfactorily described by the following equation:

$$\varepsilon_t = \frac{\varepsilon_{to}}{\ln 2} \cosh^{-1}(2^{Z\theta/\varepsilon_{to}}) \tag{6}$$

or approximated by the simple formula,

$$\varepsilon_t \approx \varepsilon_{to} + Z\theta \tag{7}$$

where Z is the Zener-Hollomon parameter (h), and ε_{to} is another creep parameter (m/m), the meaning of which is explained in Figure 2b. The Zener-Hollomon parameter is defined as (9)

$$Z = \dot{\varepsilon}_{ts} \, e^{\Delta H/RT} \tag{8}$$

where $\dot{\varepsilon}_{ts}$ is the rate of secondary creep (h^{-1}), at temperature T. The two creep parameters, Z and ε_{to}, are functions of the applied stress only (i.e., independent of the temperature).

For most materials creep becomes noticeable only if the temperature is higher than about one third of the melting temperature. The creep of concrete at room temperature is due to the presence of water in its microstructure (10). Mukaddam and Bresler (11) and Schneider (12) described procedures for correlating experimental creep data for concrete at moderately elevated and high temperatures, respectively.

THERMAL PROPERTIES

Dilatometric Curves

The dilatometric curve is a record of the fractional change of a linear dimension of a solid at steadily increasing or decreasing temperature. With mathematical symbolism, the dilatometric curve is a plot of

$$\frac{\Delta\ell}{\ell_o} \quad \text{against} \quad T$$

where $\Delta\ell = \ell - \ell_o$, and ℓ and ℓ_o are the changed and original dimensions of the solid (m), respectively, the latter usually taken at room temperature. $\Delta\ell$ reflects not only the linear expansion or shrinkage of the material, but also the dimensional effects brought on by possible physicochemical changes ("reactions" in a generalized sense).

The heating of the solid usually takes place at an agreed-upon rate, 5°C/min as a rule. Because the physicochemical changes proceed at a finite rate and some of them are irreversible, the dilatometric curves obtained by heating the material rarely coincide with those obtained during the cooling cycle. Sluggish reactions may bring about a steady rise or decline in the slope of the dilatometric curve; fast reactions may appear as discontinuities in the slope. Heating the material at a rate higher than 5°C/min usually causes the reactions to shift to higher temperatures and to develop faster.

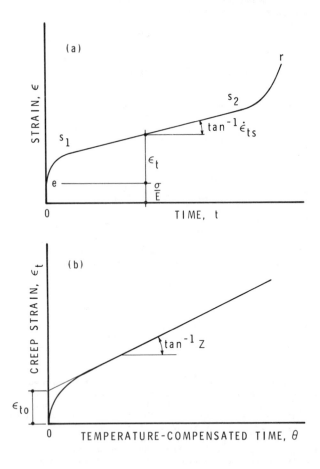

Figure 2. a) Strain–time curve; T = const, σ ≈ const
           b) Creep strain versus temperature–compensated time curve;
              σ ≈ const

Thermogravimetric Curves

The thermogravimetric curve is a record of the fractional variation of the mass of a solid at steadily increasing or decreasing temperature. Again, with mathematical symbolism, a thermogravimetric curve is a plot of

$$\frac{M}{M_o} \quad \text{against} \quad T$$

where M and M_o are the changed and original masses of the solid (kg), respectively, the latter usually taken at room temperature. If the curve is obtained by heating the solid, the agreed-upon rate of heating is, again, 5°C/min.

The thermogravimetric curves reflect reactions accompanied by loss or gain of mass but, naturally, they do not reflect changes in the materials' microstructure or their crystalline order. $M/M_o = 1$ is the dilatometric curve for a chemically inert material. Again, an increase in the rate of heating usually causes the chemical reactions to shift to higher temperatures and to develop faster.

Density and Porosity

The density, ρ (kg/m^3), in oven-dry conditions, is the mass of a unit volume of the material, comprising the solid itself and the gas-filled pores. Assuming that the material is isotropic with respect to its dilatometric behaviour, its density at any temperature can be calculated from the thermogravimetric and dilatometric curves.

$$\rho = \rho_o \frac{\dfrac{M}{M_o}_T}{\left(1 + \dfrac{\Delta \ell}{\ell_o}_T\right)^3} \tag{9}$$

where ρ_o is the density of the solid at the reference temperature (usually room temperature), and the T subscript indicates values pertaining to temperature T in the thermogravimetric and dilatometric records.

The mixture rule in its simplest form (eq. (1) with $\zeta = 1$) applies to the density of composite solids.

$$\rho = \sum_i v_i \rho_i \tag{10}$$

where the i subscript relates to information on the i-th component. If, as usual, the composition is given in mass fractions rather than volume fractions, the volume fractions are to be calculated as

$$v_i = \frac{\dfrac{w_i}{\rho_i}}{\sum_i \dfrac{w_i}{\rho_i}} \tag{11}$$

where w_i is the mass fraction of the i-th component (kg/kg).

True density, ρ_t (kg/m^3), is the density of the solid in a poreless condition. Many building materials are non-existent in such a condition, and therefore ρ_t may be a theoretical value derived on crystallographic considerations, or determined by some standard technique (e.g., ASTM C135 (13)). The relationship between the porosity and density is

$$P = \frac{\rho_t - \rho}{\rho_t} \tag{12}$$

The overall porosity of a composite material consisting of porous components is given by

$$P = \sum_i v_i P_i \tag{13}$$

where again the i subscript relates to the i-th component of the material.

Calorimetric Curves

A calorimetric curve describes the variation with temperature of the apparent specific heat of a material at constant pressure, c_p (J/kg K). If the heating of the solid is accompanied by physicochemical changes ("reactions"), the c_p becomes a function of the degree of conversion from the reactant(s) into the product(s). For any temperature range where conversion takes place (5, 14),

$$c_p = \overline{c}_p + \Delta H_p \frac{d\xi}{dT} \tag{14}$$

where $\overline{c}_p$ is the specific heat for that mixture of reactants and (solid) products that the material consists of at a given stage of the conversion characterized by the reaction progress variable, ξ (dimensionless, $0 \leqslant \xi \leqslant 1$), and ΔH_p is the latent heat associated with the conversion (J/kg).

As this equation and Figure 3 show, in temperature intervals of physicochemical instability the apparent specific heat consists of sensible

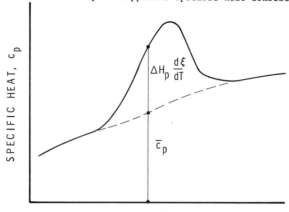

TEMPERATURE, T

Figure 3. The apparent specific heat

heat and latent heat contributions. The latter contribution will result in a peak in the calorimetric curve, maximum if the reaction is endothermic, minimum if it is exothermic.

In heat flow studies it is usually the ρc_p product rather than c_p which is needed as input information. This product is referred to as "volumetric" specific heat.

For a long time adiabatic calorimetry was the principal method for studying the shape of the c_p versus T relationship. Nowadays differential scanning calorimetry (DSC) is the most commonly used technique for mapping the curve in a single temperature sweep at a desired rate of heating. The rate of heating is, again, usually 5°C/min. At higher heating rates the peaks in the DSC curves tend to shift to higher temperatures and become sharper.

Materials that undergo exothermic reactions may yield negative values in the calorimetric curve. A negative value for c_p indicates that at the applied (and enforced) rate of heating the rate of evolution of reaction heat exceeds the rate of absorption of sensible heat by the material. In natural processes the apparent specific heat can never be negative, because the heat evolving from the reaction is either scattered to the surroundings or, if absorbed by the material, causes a very fast temperature rise. If the heat of reaction is not very high, obtaining non-negative values for c_p can be achieved by suitably raising the scanning rate.

If experimental information is not available, the c_p versus T relation can be calculated from data on heat capacity and heat of formation, tabulated in a number of handbooks (e.g., Refs. 15 and 16). The specific heat of compounds not listed in handbooks can often be estimated with the aid of the Neumann-Kopp law concerning the additivity of heat capacities. Examples of such calculations can be found in Ref. 5 where, based on handbook data, information is developed for the apparent specific heat versus temperature relation for three cement pastes and four concretes.

THERMAL CONDUCTIVITY

Heat transmission solely by conduction can take place only in poreless, non-transparent solids. In porous solids (most building materials) the mechanism of heat transmission is a combination of conduction, radiation, and convection. The thermal conductivity for such solids is, in a strict sense, merely a convenient empirical factor that makes it possible to describe the heat transmission process with the aid of the Fourier law. That empirical factor will depend on the conductivity of the solid matrix, as well as on the porosity and pore size. At higher temperatures, because of the increasing importance of radiant heat transmission through the pores, it becomes sensitive to the temperature gradient.

Clearly, measured values of the thermal conductivity depend on the temperature gradient employed in the test, and therefore great discrepancies may be found in thermal conductivity data reported by various laboratories. A thermal conductivity value yielded by a particular technique is, in a strict sense, applicable only to heat flow processes similar to that employed in that technique.

Experimental data indicate that porosity is not a greatly complicating factor as long as it is not larger than about 0.1. With insulating materials, however, the porosity may be 0.8 or higher. Conduction through the solid matrix may be quite an insignificant part of the overall heat transmission process, and therefore using the Fourier law of heat conduction in analyzing heat transmission may lead to deceptive conclusions.

If the solid is not oven-dry, any temperature gradient in it will bring about a migration of moisture, mainly by an evaporation-condensation mechanism (17). The migration of moisture is usually, but not necessarily, in the direction of heat flow, and manifests itself as an increase (or decrease) in the apparent thermal conductivity of the solid. Furthermore, even oven-dry solids may undergo decomposition (mainly dehydration) reactions at higher temperatures, and the sensible heat carried by the gaseous decomposition products as they move in the pores add to the complexity of the heat flow process. At present there is no way of satisfactorily accounting for these mass transfer processes in studies of heat flow processes occurring under fire conditions.

The thermal conductivity of layered, multiphase solid mixtures depends on whether the phases lie in the direction of or normal to the direction of heat flow. The simple mixture rule is applicable, with $\zeta = 1$ in the former case and $\zeta = -1$ in the latter. Thus, in those two limiting cases

$$k = \sum_i v_i k_i \tag{15}$$

and

$$\frac{1}{k} = \sum_i \frac{v_i}{k_i} \tag{16}$$

where k and k_i are the thermal conductivities of the mixture and its constituents (W/mK), respectively.

There is substantial evidence (3, 18) that eqs. (2) and (3) with $n = 3$, are applicable to any two-phase material consisting of a continuous and a discontinuous solid phase, irrespective of the geometry of the discontinuous phase.

$$k = \frac{v_1 k_1 + \alpha \, v_2 k_2}{v_1 + \alpha \, k_2} \tag{17}$$

where

$$\alpha = \frac{3k_1}{2k_1 + k_2} \tag{18}$$

and the subscripts 1 and 2 relate to the continuous and discontinuous phases, respectively. When both phases are essentially continuous, as with most porous materials (with air as the second phase), a lower value of n seems to be applicable (5): about $n = 1.5$. If the conductivity of air is negligibly small in comparison with that of the solid, the following is a good approximation:

$$k = k_s(1 - P) \tag{19}$$

where k_s is the conductivity of the solid itself (W/mK). At higher temperatures, however, the pores become conductive owing to radiation. The thermal conductivity is customarily expressed as a sum of two terms,

$$k = k_s(1 - P) + k_R \tag{20}$$

where

$$k_R = 4\sigma F d_p T^3 \tag{21}$$

and k_R is the so-called radiant conductivity (W/mK), σ is the Stefan-Boltzmann constant (W/m^{2}K^4), d_p is the characteristic pore size (m), and F is a constant characteristic of the material and the pore geometry (dimensionless).

The thermal conductivity of solids is a structure-sensitive property. If a solid is crystalline, its conductivity is relatively high at room temperature but gradually decreases as the temperature rises. If, on the other hand, it is predominantly amorphous, its conductivity is low at room temperature and increases slightly with temperature. The conductivity of porous crystalline materials may also increase at very high temperatures, owing to the radiant conductivity of the pores.

Unfortunately, to the writer's knowledge, no scanning technique is available for the acquisition of a continuous thermal conductivity versus temperature curve from a single test. Such a curve must be estimated from the results of tests conducted at various temperature levels. Special problems arise in the temperature intervals of physicochemical instability. Both the steady-state and variable-state techniques of measuring thermal conductivity require the stabilization of either a temperature level or a temperature distribution, and thereby a certain microstructural condition in the specimen prior to the test. The test results can therefore be viewed as points on a continuous thermal conductivity versus temperature curve obtained by an imaginary scanning technique performed at an extremely slow scanning rate. Since each point pertains to a more or less stabilized microstructural pattern, there is no way of knowing how the thermal conductivity would vary during the course of a physicochemical process developing at a finite rate and varying microstructure.

Owing to non-reversible microstructural changes brought about by heating, the thermal conductivity of building materials, and perhaps most other materials, is usually different for the heating and cooling cycles.

SOURCES OF INFORMATION

Information on the properties of building materials at elevated temperatures is scattered throughout the literature. There are a few publications, however, which contain information in a way particularly suitable for use by fire safety specialists. One such publication (19) is a source of information on the thermal properties of 31 building materials in Groups L/I and I. Another comprehensive source of information, specifically on concrete and steel, deals with their fire endurance (20). Two other reports give a wealth of information on the behaviour of steel (21) and concrete (22) at high temperatures.

Among many other publications, a few, dealing in more detail with specific material properties at elevated temperatures, are mentioned below:

Steel, mechanical properties:

Stress-strain curves for a structural steel and a prestressing wire (23)
Modulus of elasticity (24, 25)
Creep (23, 26, 27)

Concrete, mechanical properties:

Stress-strain curves (28, 29, 30)
Modulus of elasticity (28, 31, 32)
Strength (28, 32, 33, 34, 35, 36, 37)
Creep (12, 38)

Concrete, thermal properties:

Dilatometric and thermogravimetric curves (14)
Apparent specific heat (5, 19)
Thermal conductivity (14, 19)
Thermal conductivity of about 50 rocks (potential concrete aggregates) (39)

Wood, mechanical properties:

Modulus of elasticity, compressive strength (40, 41, 42)

Gypsum, thermal properties:

Volumetric specific heat (43)

NOMENCLATURE

c	specific heat, J/kgK
$\bar{c}$	specific heat for a mixture of reactants and solid products, J/kgK
d_p	characteristic pore size, m
E	modulus of elasticity, MPa
F	material constant, dimensionless
ΔH	activation energy for creep, J/kmole
ΔH_p	latent heat (at constant temperature) associated with a conversion, J/kg
k	thermal conductivity, W/mK
ℓ	dimension, m
$\Delta \ell$	$= \ell - \ell_o$
M	mass, kg
n	material constant, dimensionless
P	porosity, m^3/m^3
R	gas constant, J/(kmole)K
S	specific surface, m^2/m^3
t	time, h

T	temperature, K (or °C)
v	volume fraction, m^3/m^3
w	mass fraction, kg/kg
Z	Zener-Hollomon parameter, h^{-1}
α	expression defined by eq. (3), dimensionless
ϵ,	strain (deformation), m/m
ϵ_{to}	creep parameter, m/m
$\dot{\epsilon}_{ts}$	rate of secondary creep, h^{-1}
θ	temperature-compensated time, h
ξ	reaction progress variable, dimensionless
π	material property (any)
ρ	density, kg/m^3
ρ_t	true density, kg/m^3
σ	stress, MPa
σ	Stefan-Boltzmann constant, W/m^2K^4

SUBSCRIPTS

i	of the i-th constituent
o	original value at reference temperature
s	of the solid matrix
p	at constant pressure
t	true
T	at temperature T
u	ultimate
y	yield

REFERENCES

1. Harmathy, T.Z., Moisture sorption of building materials. Techn. Paper No. 242, NRCC 9492, Div. Bldg. Res., Nat. Res. Counc., Ottawa, 1967.

2. Bruggeman, D.A.G., Uber die geltungsbereiche und die Konstantenwerte der verschiedenen Mischkorperformeln Lichteneckers. Phys. Zeitschrift, 37, 1936, 906.

3. Hamilton, R.L.; Crosser, O.K., Thermal conductivity of heterogeneous two-component systems. I & EC Fundamentals, 7, 1962, 187.

4. Maxwell, J.C., A Treatise on Electricity and Magnestism. Vol. I, 3rd ed., Dover, New York, 1954, 435.

5. Harmathy, T.Z., Thermal properties of concrete at elevated temperatures. J. of Matls., 5, 1970, 47.

6. Dorn, J.E., Some fundamental experiments on high-temperature creep. J. of Mech. Phys. Solids, 3, 1954, 85.

7. Harmathy, T.Z., Deflection and failure of steel-supported floors and beams in fire. Fire Test Methods - Restraint and Smoke, ASTM STP 422, 1967, 40.

8. Harmathy, T.Z., A comprehensive creep model. Transactions, ASME J. Basic Engng., 89, 1967, 496.

9. Zener, C.; Holloman, J.H., Effect of strain rate on the plastic flow of steel. J. Appl. Phys., 15, 1944, 22.

10. Ramachandran, V.S.; Feldman, R.F.; Beaudoin, J.J., Concrete Science. Heyden & Son Ltd., London, 1981, 40.

11. Mukaddam, M.A.; Bresler, B., Behavior of concrete under variable temperature and loading. Concrete for Nuclear Reactors, ACI SP-34, 1972, 771.

12. Schneider, U., Ein Beitrag zur Frage des Kriechens und der Relaxation von Beton under hohen Temperaturen. Institut fur Baustoffe, Techn. Univ. Braunschweig, Heft 42, 1979.

13. C135 Test Method for True Specific Gravity of Refractory Materials by Water Immersion. Annual Book of ASTM Standards, Vol. 15.01.

14. Harmathy, T.Z.; Allen, L.W., Thermal properties of selected masonry unit concretes. J. ACI, 70, 1973, 132.

15. Chemical Engineers' Handbook. Ed. J.H. Perry, 3rd edition, McGraw-Hill Book Co., Inc., 1950.

16. Eitel, W., Thermochemical Methods in Silicate Investigation, Rutgers Univ. Press, New Brunswick, NJ, 1952.

17. Harmathy, T.Z., Simultaneous moisture and heat transfer in porous systems, with particular reference to drying. I & EC Fundamentals, 8, 1969, 92.

18. DeVries, D.A., The thermal conductivity of granular materials. Bulletin, Inst. Inerntl. Froid, Annexe 1952-1, 115.

19. Harmathy, T.Z., Properties of building materials at elevated temperatures. DBR Paper No. 1080, NRCC 20956, Div. Bldg. Res., Nat. Res. Counc., Ottawa, 1983.

20. Guide for Determining the Fire Endurance of Concrete Elements. ACI 216R-81, 1982.

21. Anderberg, Y., Behaviour of steel at high temperatures. Preliminary Rpt., RILEM Cte. 44-PHT, 1983.

22. Properties of Materials at High Temperatures - Concrete. Ed. U. Schneider, Dept. Civil Engng., Kassel, Germany, 1985.

23. Harmathy, T.Z.; Stanzak, W.W., Elevated-temperature tensile and creep properties of some structural and prestressing steels. Fire Test Performance, ASTM STP 464, 1970, 186.

24. European Recommendations for the Fire Safety of Steel Structures. European Convention for Constructional Steelwork (ECCS), Techn. Cte. 3, Fire Safety of Steel Structures, Elsevier Sci. Publ. Co., New York, 1983, 106.

25. Anderberg, Y., Mechanical properties of reinforcing steel at elevated temperatures (in Swedish). Tekniska meddelande nr. 36, Halmsted Jarnverk AB, Lund, 1978.

26. Knight, D.C.; Skinner, D.H.; Lay, M.G., Prediction of isothermal creep. Rept. MRL 18/2, Melbourne Labs., Clayton, Australia, 1971.

27. Williams-Leir, G., Creep of structural steel in fire: Analytical expressions. Fire & Matls., 7, 1983, 73.

28. Harmathy, T.Z.; Berndt, J.E., Hydrated portland cement and lightweight concrete at elevated temperatures. J. ACI, 63, 1966, 93.

29. Schneider, U., Behaviour of concrete under steady state and non-steady state conditions. Fire & Matls., 1, 1976, 103.

30. Anderberg, Y.; Thelandersson, S., Stress and deformation characteristics of concrete at high temperatures, 2. Experimental investigation and material behaviour model. Div. of Struct. Mech. Concr. Constr., Lund Inst. Technol., Bull. 54, 1976.

31. Cruz, C.R., Elastic properties of concrete at high temperatures. J. PCA Res. Devel. Labs., 8, 1966, 37.

32. Saemann, J.G., and Washa, G.W., Variation of mortar and concrete properties with temperature. J. ACI, 54, 1957, 385.

33. Abrams, M.S., Compressive strength of concrete at temperatures to 1600°F. Temperature and Concrete, ACI SP-25, 1971, 33.

34. Zoldners, N.G., Effect of high temperatures on concrete incorporating different aggregates. Mines Branch Res. Rep. No. 64, Dept. Mines Techn. Surveys, Ottawa, 1960.

35. Malhotra, H.L., The effect of temperature on the compressive strength of concrete. Mag. Concr. Res. (London), 8, 1956, 85.

36. Binner, C.R.; Wilkie, C.B.; Miller, P., Heat testing of high density concrete. Declassified AEC Report No. HKF-1, U.S. Atom. En. Comm., Washington, DC, 1949.

37. Weigler, H. and Fischer, R., Uber den Einfluss von Temperaturen uber 100°C auf die Druckfestigkeit von Zementmortel. Bull. No. 164, Deut. Aussch. Stahlbeton, Berlin, 1964.

38. Cruz, C.R., Apparatus for measuring creep of concrete at high temperatures. J. PCA Res. Devel. Labs., 10, 1968, 36.

39. Birch, F.; Clark, H., Thermal conductivity of rocks and its dependence upon temperature. Am. J. Sci., 238, 1940, 542.

40. Gerhards, C.C., Effect of moisture content and temperature on mechanical properties of wood: An analysis of immediate effects. Wood & Fiber, 14, 1981, 4.

41. Schaffer, E.L., State of structural timber fire endurance. Wood & Fiber, 9, 1977, 145.

43. Schaffer, E.L., Structural fire design: Wood. Res. Paper FPL 450, U.S. Dept. Agric., Forest Sci., Forest Prods. Lab., Madison, WI, 1984.

43. Harmathy, T.Z., A treatise on theoretical fire endurance rating. Fire Test Methods, ASTM STP 301, 1961, 10.

DESIGN PHILOSOPHY

R.D. ANCHOR
Consulting Engineer and
Department of Civil Engineering
University of Aston in Birmingham

ABSTRACT

The paper considers the design of structures against fire from the viewpoint of the practising engineer and discusses a number of elements in design which require further investigation or consideration.

INTRODUCTION

The subject of structural design against fire brings together contractors, practising engineers and architects, research workers and scientists. Each has a contribution to make to the advancement of knowledge and practice, but it is commonplace that it is very easy to disregard work outside the area of one's own expertise. Indeed, it is probable that ideas put forward by one practitioner will be aiming at a different objective from others in the field. A conference of the various interests presents an opportunity for ideas to be exchanged and hopefully some agreement to be reached for further progress. The title of my paper poses the need for two definitions. "Design" has been well described in a document published by The Institution of Structural Engineers[1], London as being "governed by the general desire to have what is needed at minimum cost; it is an art concerned with the adequate" and "Philosophy" which is defined in the dictionary as "love of wisdom or knowledge especially that which deals with <u>ultimate reality</u>". As a practicing engineer, the last two words particularly appeal to me in that whatever theories and/or experiments are undertaken, unless they relate to what is appropriate for designers to use in their every-day work they will lead to a dead-end.

The following paragraphs discuss some of the issues

facing designers at the present time.

REGULATIONS

In the U.K., designers are required to work in
accordance with the current Building Regulations[2] published
by the Government and operated by the Local Authorities.
There has recently (November 1985) been published a
completely rewritten set of Regulations in which all the
detailed information and clauses used by the designer have
been removed from the main body of the Regulations and
placed in "Approved Documents". This is considered to be a
simplification and to be more flexible, in that new approved
documents can easily be added to the list without altering
the Regulations themselves. The basis of the Regulations
remains the same. The intention is to provide for the
health, safety and convenience of the population when using
any type of building. This objective may be questioned as
being too limited in modern conditions. Whilst safety of
people is of paramount importance, it may be argued that as
far as structural fire resistance is concerned (i.e. not
smoke control) we are largely successful in ensuring that a
building will not collapse until all the occupants have been
evacuated. What we now need is to extend our objective to
include the desirability of limiting the spread of fire
within a building so that the building structure itself and
the contents of the building are more likely to be saved. A
typical simple example is a small separate single-storey
building which houses a computer installation. The value of
the building may be £25,000, the value of the contents
£5,000,000 and it may be occupied by only six people. The
current regulations are irrelevant to this situation. It
may be argued that the insurance market rather than the
Government should be involved in considerations of saving
property and contents. However, there is generally no
contact between the insurers of a building and the
professional designers. Furthermore, the scales of premiums
used by insurance companies do not relate to Structural
Codes of Practice but are based on rules laid down by the
Insurers as a result of their experience. It seems curious
that the insurance world is so divorced from the discussions
which take place in the professions. More specifically,
insurers do not contribute to the synthesis leading to a
practical design. The Author is not able to suggest how
this state of affairs could be altered. A further
possibility is for architects and others to educate building
owners in the subject. In the Merrett-Cyriax report[3] it is
stated that an appreciable number of industrial firms
experiencing a large fire in their premises, go out of
business altogether. Although it is no less likely for a
fire to start in a modern building,

as compared with an older property, the consequences of the fire will be more profound in the older property which is unlikely to have a framed structure and will therefore collapse much more quickly leading to increased danger to the fire-fighters.

In the Merrett Cyriax[3] report it is also stated that 77% of all fire damaged buildings investigated were rebuilt to the original specification. It appears that Insurance Companies and Industrialists need further education in the subject of fire protection to avoid a continuingly unsatisfactory situation.

Referring again to the Building Regulations[2], it was to be expected that the new Regulations would accord "Approved Document" status to BS 8110 Parts 1[4] and 2[5] but this has not happened in respect of structural design against fire. The design method in BS 8110 part 2[5] which enables a numerate design to be prepared for reinforced concrete beams and slabs allowing for continuity was written after some years of deliberation and discussion. It is now unsatisfactorily vague as to whether the BS 8110 part 2 clauses on design will generally be acceptable to local Building Surveyors.

FIRE LOAD

The U.K. regulations divide all buildings into one of eight purpose groups. The prescribed period of fire resistance depends on the purpose group, the size of the fire compartment and whether the section of the structure contains a basement. This generally provides a simple and satisfactory system of classification for smaller buildings, but where a building is used for storage (including certain shop premises), there is merit in considering the actual fire load likely to be available, and designing the structure accordingly. Depending whether the contents are (say) steel or a highly combustible material the structural fire design could then be based on a logically determined period of fire resistance rather than a standard period. The difficulty with this approach is the possibility of a future change in use of the building which could alter the fire load. Any consequent necessary upgrading of the structural fire protection could be relatively expensive. It has been said that a building should never be designed for a particular use but only in a generalised way, and the Author has some sympathy with this view.

STRUCTURAL DESIGN

The choice of structural material is perhaps the most fundamental decision which the designer has to make. There is usually a broad consensus amongst designers at a given time as to the appropriate material to use in particular circumstances, but over a longer period of time, trends are discernible. At present in the U.K., structural steel frames for office type buildings with or without composite concrete/profiled steel sheet floors are becoming more common. The choice is made on grounds of suitability for purpose, span, load-carrying capacity and cost. Factors such as speed of erection may also be relevant. Fire resistance is probably usually considered as an item included in the total cost as it is technically always possible to provide the appropriate degree of fire resistance to any structure. However, although the choice of structural material is not made on grounds of fire resistance, the amount of work required to provide the required fire resistance will depend very much on the material.

It is usual to consider fire resistance in terms of individual structural elements although in BS 8110 part 2[5] a method of calculation for the fire resistance of concrete beams taking into account structural continuity is specified. This method will be considered in more detail by other authors. It may eventually be possible to allow for the effect of membrane forces where these add to structural stability in fire.

MAINTENANCE

An aspect of structural fire resistance which is largely disregarded is the need to maintain the integrity of the original fire design. Where suspended ceilings are part of the fire protection to the structure, any maintenance work on the services must ensure that the fire protection is reinstated effectively. Equally, the original fire stopping of vertical or horizontal service runs through compartment floors or walls must be replaced when the services are altered or renewed. Similar problems occur when hollow protection is used for steelwork which may be damaged either by accidental impact or by maintenance workers. Intumescent paint may be degraded by subsequent damage.

WORKMANSHIP

The design of any structure should always be made with a view to ease of construction. Some fire resistance techniques are particularly susceptible to accuracy of workmanship; fire stopping at compartment boundaries has already been mentioned. Another problem occurs where extra mesh reinforcement is required in the concrete cover of reinforced concrete columns and beams. Theoretically, the extra mesh should be fixed in the mid-depth of the cover but in practice this is nearly impossible to achieve, and is clearly in contravention of the requirements for durability. There is therefore a conflict between one requirement and another.

Workmanship is also related to supervision of construction in that unless the detailed supervision is adequate, the quality of workmanship will tend to fall.

DETAILING

Detailing is particularly important in structural concrete construction but is also relevant in other forms of structure. It is possible to completely change the effective fire resistance of a reinforced concrete beam from say 4 hours to 15 minutes by poor or uninformed detailing. A recent contract for a steel portal frame structure required 4 hours fire resistance for the perimeter walls. The steel stanchions were protected with brickwork casings, and the steel beams giving support to the wall panels were protected with an applied insulation material. The specification required the insulation to be stuck to the brickwork in order to cover the steel. This was queried by Building Control authorities and it transpired that to give 4 hours protection, the insulation needed to be screwed to the brickwork. This is the type of detail which is often overlooked.

CONCLUSIONS

The present state of fire design is interesting, in that over the last ten years considerable progress has been made in identifying the factors which are relevant to improved design. Further research is now required, particularly in the area of structural continuity, where a larger test facility is necessary, together with a greater effort to persuade practitioners of the need for clear thinking when preparing fire designs.

110

REFERENCES

1. The Institution of Structural Engineers, Aims of Structural Design, London 1969

2. Her Majesty's Stationery Office, The Building Regulations 1985, London 1985

3. Merrett Cyriax Associates, The Role of Structural Fire Protection, London 1969

4. British Standards Institution, Structural Use of Concrete, Part 1 Code of Practice for Design and Construction, London 1985

5. British Standards Institution, Structural Use of Concrete Part 2 Code of Practice for Special Circumstances, London 1985

THE WHOLE STRUCTURE

FORREST, JOHN C.M.
Kenchington Little & Partners
Building & Civil Engineering Consultants
London U.K.

INTRODUCTION

The ten years that have elapsed since the first Aston International Conference have seen significant developments in the form that structures are taking to meet architectural designs.

In multi-storey office blocks, hotels and shopping centres the use of Atria to enclose space for environmental comfort and aesthetic appeal has brought new thinking into the control of smoke dispersion, escape facilities and access for fire fighting.

In industrial buildings there has been rapid development of the use of high technology buildings. This is a recent term given to new buildings incorporating aesthetic appeal with very high standards of environmental and communication services, and flexible use of space. Many such structures gain aesthetic appeal from the use of an exposed structural form differing greatly from the traditional column, beam and slab approach.

In cities the high cost of land, lack of space for construction and inherent usefulness of existing structures have led to major refurbishment of older buildings. Many such structures are being revitalised by alterations to special layout, communications between floors, addition of complex services e.g. air conditioning, computer facilities and the like.

These developments were little known ten years ago – now they are part of most developers thinking and have permeated through all the design professions.

This paper refers to the design of passive resistance to fire of whole structures (as opposed to buildings) and therefore excludes reference to active measures for fire control such as sprinklers or fire suppresive gases which are part of building design.

OBJECTIVES IN DESIGN

The approach to the fire design of the whole structure contains fundamental objectives which have to be met irrespective of the type of building. The two main objectives are safety of life and safety of property. Each of these can be broadly sub-divided into two.

Safety of Life	Safety of Property
. The occupants of the building	. The Building
. The fire fighters	. The Contents

To meet these two objectives the designer has to understand the behaviour of the whole structure under fire attack from any selected quarter.

The designer does not achieve these objectives by mere compliance with Regulations, Building Codes and Insurance Construction Rules. Such documents are usually related only to the behaviour of single elements of structure as obtained from fire tests in research furnaces or historical evidence from fire damage surveys. They do not provide compliance criteria for assemblies of two or more elements together, nor for the interaction throughout the building. The designer has to assess such aspects separately and take due account in the design.

Fortunately, the information documented by many fire specialists around the world now provides a reasonable data base on which design checks for the fire integrity of whole buildings can be based.

WHOLE STRUCTURE BEHAVIOUR IN ACTUAL FIRES

Over the last 25 years there have been a number of notable fires in buildings from which the behaviour of the whole structure can be assessed.

Apart from considerations of the fire behaviour of single elements the following factors need to be taken into account in assessing the behaviour of whole structures in fire.

. Continuity within the structure
. Retention of compartmentation
. Provision for Expansion
. Load Re-distribution
. Maintenance of Stability
. Type of Structure and Materials used

In many instances of fires these factors overlap in their effects on the whole structure.

Each of these factors is examined by reference to reports of notable fires which have revealed strengths and weaknesses in the design or construction.

CONTINUITY WITHIN THE STRUCTURE

Experience from reported fires has demonstrated the beneficial effects complete continuity gives to a whole structure.

Many examples exist of continuous structures where extensive fire attack has not damaged the building beyond repair. Indeed the corollary may be said to be true i.e. that where continuity of structural frame is broken there is a risk in fire that irreparable damage may be done. For instance, it is not easy to make precast concrete floors fully continuous in action with a steel framework whereas, an insitu concrete floor cast on permanent steel formwork can be made fully continuous with a steel frame in modern composite construction techniques.

EXAMPLE - SHOP/CAR PARK, WITTERSWAND, SOUTH AFRICA - MARCH 1981

A four storey building with ground floor used as retail shopping, 60 m long x 23.6 m wide, 1st floor as offices and the other two floors used as a car park. The long walls were of masonry with built-in columns, an intermediate row of columns at 6.5 m centre were topped by, main beams, 20.50 mm x 600 mm deep, reinforced with 16 mm deformed high strength bars. The 170 mm in situ concrete floor (70 mm screed), was provided with 600 mm deep x 300 mm wide ribs at 1250 mm centre. The ribs were formed by using 300 x 100 mm deep prestressed concrete planks with one end resting on the walls and the other forming the lower corner of the main beam but without any steel ties. The pretensioning was provided by two 12.7 mm strands with 25 mm soffit cover. Cross beams were provided at 3 m centres, reinforced with 16 mm deformed high tensile bars with 110 mm cover. On the top of the slab brick walls were erected to provide division between offices.

The fire started at the rear of the shop in unpacked goods and spread 2/3rds of the distance to the front of the shop. It was controlled in about 1½ hours but was not completely extinguished for another 5 hours. It did not spread to any other part of the building. (See Figure 1.)

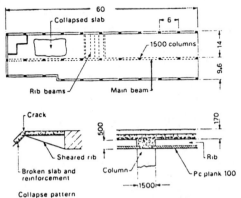

Figure 1. General arrangement and collapse pattern.

No collapse had occurred during the fire but a part of the floor measuring approximately 20 x 12 m, collapsed some time during 48 hours after the fire. Shear failure of the ribs occurred, near the support to the main beam and the brick wall, with a clean break of the reinforcing mesh. No shear reinforcement had been provided in the ribs. Where the collapse occurred the fire was considered to have been severe. Samples of steel bars from the cross beams in the damaged area showed no loss of yield or the ultimate strength. The stranded cables from the same area showed that their rupture strength had been reduced to 84 per cent, with signs of yield at 66 per cent of the rupture strength. Considerable amounts of spalling from the sides of the ribs and some parts of the soffit of rc beams was observed. The plastered columns did not show any sign of damage.

The design of the floor was unsatisfactory for fire resistance, the absence of shear reinforcement in the ribs and the poor detailing of the junction of prestressed planks and the main beams would have been mainly responsible for the failure.

RETENTION OF COMPARTMENTATION

Building Codes and Regulations will determine the needs of compartmentation to suit the fire resistance periods of the building. The designer has to assess how compartments will be formed, what materials will be used and provision for horizontal and vertical access for people and goods. Experience has shown that fires will always find the weak points in compartment constructions. Usually any one of the three factors below can lead to unsuccessful compartmentation.

* Failure by insufficient tying between structural elements
* Failure by loss of support of one element
* Failure by a 'loophole' in fire integrity.

Successful compartmentation under fire attack results from design and construction which guards against the effects of these three factors considered together.

EXAMPLE - 14 STOREY OFFICE BLOCK AT ROTTERDAM NETHERLANDS - 1978

A 14-storey building, consisting of reinforced concrete columns and beams with reinforced concrete slabs, 120 mm thick, giving a plan area of 18 x 24.6 m. The columns were arranged in a grid, with spacings of 5.5 or 6.8 m, thus giving floor slabs with two different spans supported on beams on four sides. The internal beams measured 400 x 500 mm and columns were 700 x 700 mm decreasing to 400 x 400 at the top floor. The building was used as an office and the fire load has been estimated at 30 to 60 kg wood/m^2. There was a fibreboard ceiling below the concrete slab on most floors.

The fire started on the 9th floor and progressively spread upwards involving all the floors except the topmost one which was set back. Although the fire was fought by the brigade it seems that the majority of the contents in the floors involved had burnt themselves out. (See figure 2.)

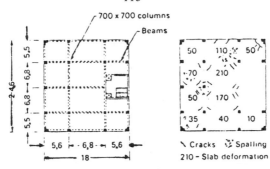

Figure 2. Layout of 10th floor and damage.

At the end of the fire the building was still standing, but spalling of concrete on the soffit of slabs, beams and column faces was observed in about 20 per cent of the exposed elements. Most of the slabs had suffered spalling along the beams sides and corners in the predominantly compressive zone exposed directly to heating. Some of the columns had suffered diagonal cracking and cracking was also noticed in floor slabs. The most prominent feature of the fire was the residual deformation of the slabs and beams. The central deformation of floors varied from 15 to 210 mm, the maximum deformation was found on the 10th storey in the floor spanning 6.8 m. The maximum deformation suffered by a beam was 45 mm, in the vicinity of the worst floor.

The prominent features of this fire were the spalling in the exposed floor slab concrete in the negative moment zone, cracks showing the stress transformation boundaries and excessive residual deformation.

PROVISION FOR EXPANSION

The very large temperature rises in structures as a result of a fire lead to large expansive forces which cause parts of the building to move many times more than the normally designed provision for seasonal movement. In movement sensitive structures the designer has therefore to assess the likely range of temperature movements occasioned by a fire and plan accordingly.

In such structures the application of heat to the underside of compartment floors will lead to high horizontal movements which must be accommodated by sufficiently flexible columns and walls. Plan configurations for multi-storey structures under 30 m square or rectangles of not more than 75 m and width not greater than 30 m have performed well. However buildings of large plan area and of 2 to 4 storeys have suffered severe damage from expansion during fire where provision for large horizontal movements was not designed. Such buildings as shopping centres, factories and warehouses have been found to be particularly susceptible.

EXAMPLE - 3 STOREY WAREHOUSE AT GHENT, BELGIUM - 1974

A 50 x 50 m square, three storey building consisting of a framework of beams and columns with ribbed in situ floor, bonded together without any expansion joints. The building was intended as a warehouse to carry a live load of 13 KN/m^2. (See figure 3.)

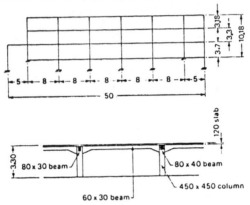

Figure 3. Structural framework and internal detail.

At the time of the fire the 2nd and 3rd floors were probably empty. The ground floor had an estimated fire load of cotton, paper rolls and viscose fibres totalling about 1,000 tonnes. About 75 minutes after the start of the fire, one corner of the building collapsed, followed by others.

At the end of the fire, one half of the building had completely collapsed and the remainder badly damaged. The damage was initially caused by the expansion of the 50 m slab without expansion joints, creating high horizontal stresses and displacements on the tops of columns causing their fracture. High bending moments were induced over the supports and it is likely that hogging occurred due to the absence of live load on the floor.

The absence of expansion joints and the deficiency of the reinforcement in the upper zone are considered to have been the most critical factors.

LOAD RE-DISTRIBUTION

The experience gained from actual fires shows that many structures under fire attack are capable of re-distribution of load from that provided in the original design. Such re-distribution is normally noted in horizontal planes where heat effects are greatest. Where load re-distribution is not possible under prolonged fire attack, local collapse of the structure has occurred.

EXAMPLE - WAREHOUSE AT MANCHESTER U.K. - OCTOBER 1967

This reinforced-concrete framed building was used as a store for
packing material, waxed paper and paraffin wax. The cover to the beam
steel was 40 mm and to the slab steel was 20 mm. The fire brought the
floors near to collapse after a period of 2 hours.

Figs 4 and 5 show to some extent the way in which the floor was
prevented from collapsing completely. The slab would appear to be acting
as a two-way catenary, the reinforcement obviously being adequately
anchored to the surrounding and adjacent beams and slabs. Without this
adequate anchorage the floor would have collapsed into the storey below.

Figure 4. View of floor in storey above the fire showing the fully
developed suspension of the slab

Figure 5. View of floor over fire
(Note exposed reinforcement acting as a catenary)

MAINTENANCE OF STABILITY

The analysis of fires where structural collapse has occurred indicates loss of stability as the main cause. In highly redundant structures loss of overall stability rarely occurs. In statically determinate structures, however, the combination of loss of secondary structural members coupled with high movements in primary members lead to unacceptable eccentricities of loading and eventual collapse.

Single storey structures are particularly vulnerable when collapse of the roof members during the fire can result in collapse of external walls from loss of top restraint.

EXAMPLE - 3 STOREY COMPLEX AT MITCHAM, SURREY, UK - DECEMBER 1969

On 6 December 1969 a fire was discovered which spread from a timber building into the 3-storey reinforced-concrete building used as a store for plastics material stored in vertical racking. In this structure stability was lost, and part of the 3-storey building collapsed, trapping and injuring several firemen.

Both suspended floors were 150 mm thick and were composed of precast reinforced-concrete T-beams supporting hollow-pot floors with topping. The first floor beams were 300 mm wide with 300 mm downstand below floor. The 2nd floor beams were also 300 mm wide with 230 mm downstand. The columns, ground to first, were 300 mm square with four 16 mm diameter bars, 50 mm cover and 6 mm links at 190 mm centre to centre. The length of column was approximately 4.3 m. The columns from first to second floor were 230 mm square with four 12 mm diameter bars, links at 150 mm centre to centre.

It is considered that one or more columns between ground and first floor on grid line D (See Figs 6 and 7) first collapsed causing the further collapse of a large area of the building. The collapse is considered to be due to:

(i)   a fire greater than one of standard intensity
(ii)   the effect of heat and fire on one side only of the column
(iii)   possible local weaknesses in the reinforced-concrete columns
(iv)   an apparent lack of transverse stiffness in the
        reinforced-concrete
        frames and floors
(v)   short curtailment of all top reinforcement.

Examination of the reinforcement and concrete indicated that the steel was not subjected to a temperature of more than 500°C for any significant length of time and that the concrete was no better than the average 1:2:4 mix.

The building was six years old at the time of the fire.

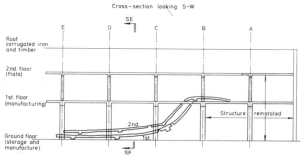

Figure 6. Sketch indicating general position of floors after collapse.

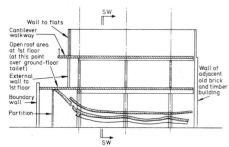

Figure 7. Sketch indicating general postion of floors after collapse.

TYPE OF STRUCTURE AND MATERIALS USED

The large majority of framed buildings constructed since the war have floors of concrete, irrespective of the material used for the frame, whether structural steel or concrete.

Steel structures cased in concrete or clad with mineral fibre fire protection have been as successful as reinforced in-situ concrete in resisting prolonged fire attack.

Structures that by choice of material or construction have not been quite so successful are those where lack of adequate continuity existed or where system building allows loopholes in defences against fire – particuarly at connection details.

EXAMPLE - 13 STOREY OFFICE BLOCK - LONDON U.K. - MARCH 1980

A 13 storey office block of rectangular plan, consisting of concrete floors, supported by edge beams resting on columns. Two rows of 300 x 600 mm columns spaced 6 m apart with a 1.5 m wide beam on top provided the central spine, and on each long face 200 mm square external mullions at 150 mm centres supported edge beams. The floor consisted of in situ reinforced concrete ribs, running transversely to the building and resting on the beams. The ribs were located to provide space for 300 mm wide hollow clay blocks, clay slip tiles were provided under the ribs which were reinforced with 2 - 19 mm twisted high yield bars. The total depth of the floor was 228 mm and the soffit of the floor as well as the central beam was plastered. A mineral fibre suspended ceiling was located 100 mm below the soffit. Two concrete cross walls, designed as fire barriers divided each floor into three parts. (See figure 8.)

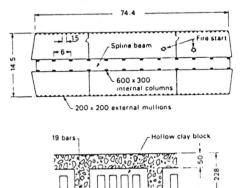

Figure 8.   Layout and floor slab details.

The fire started in two placed on the 9th floor, and because of the prevailing wind conditions flames were elongated on one side. The fire spread externally to all the higher floors in the central and the southern part containing the lift lobby. Virtually all of the combustible contents on 9 to 11th floors were consumed with minor damage to the contents of the restaurant on 13th floor.

The suspended ceiling was not of a fire resisting construction and it collapsed early in the fire. There was local damage to hollow clay blocks with the spalling of the lower face and some of the slip tiles had come down. There was no noticeable deformation of the floors and they were considered to be structurally sound. The inner corners of the facade mullions on the leeward side had suffered spalling at the arises but had not collapsed. The internal columns had not suffered damage due to protection provided by the plaster finish.

The suspended ceiling and the plaster finish provided a degree of protection to the floor and the internal beams and limited the extent of damage. The spalling of concrete corners of external mullions was worsened by the weather conditions. The composite nature of the construction and good detailing prevented any serious structural damage.

THE DEVELOPMENT OF FIRE TEST PROCEDURES

Over the last ten years there has been a much better understanding of the requirements of a fire test on a structural element. Previously fire tests had been conducted on the basis of how long the unit withstood the effects of the 'standard' fire and what temperatures were reached by the steel/reinforcement/prestressing strand, etc within the structure under full working load. Without detriment to the research workers involved, insufficient attention was given to the support of the test specimen during the fire test and where thought was given the design of early test furnaces denied any method of support beyond the simple supported condition.

However, the construction of new furnaces in most developed countries has now altered the test scenario significantly. Today it is possible to mount most test specimens in a test rig over or within a furnace and apply external loads so that a much better relationship to the behaviour of that element in a real fire can be achieved. The results of such a fire test are therefore more meaningful and capable of introduction into building codes with greater assurance of meeting safety of life objectives.

Recent developments at the Fire Insurers Research and Testing Organisation (FIRTO) Laboratories at Borehamwood, Herts. have enabled for instance, the testing of two-way spanning slabs of 'waffle' configuration as closely representing a section of the floor from an actual design for a building. The results of these tests carried out in 1985 by the Construction Industry Research and Information Association (C.I.R.I.A.) have led to a better understanding of the efforts of spalling and positioning of reinforcement within such slabs to resist fire attack.

However big the furnace and scope of its facilities to apply external loads to simulate actual forces in structures it is still impractical to attempt full-scale fire tests on whole structures. Structures of the proportion of a single detached house have been tested in several countries in Europe but multi-storey structures, large development structures (e.g. shopping centres), high technology structures are clearly beyond the scope of any full-scale test facilities. In 1986 the designer is therefore faced with the knowledge that designs can proceed on the fire resistance of individual elements tested to accord with their anticipated behaviour within a future structure, backed up with the past experience recorded from the behaviour of whole structures under fire attack. In the opinion of the author this approach is the best that can be attainable until full mathematical modelling of whole structures under fire attack is accepted into structural engineering design and practice.

ATTENTION TO DETAIL

The experience gained from previous fires indicates that close attention to correct detailing of structural elements will help to ensure adequate fire resistance for whole structures. Some fundamental questions on the lines of those below should be answered by the designer.

. Can the building expand horizontally?
. Can columns and walls flex adequately in the desired direction?
. Can integrity of compartments be maintained?
. Can load be redistributed via arching/suspension systems?
. Can stability be maintained?
. How will joints in the structure perform?
. What services penetrate the structure?
. What other aspects affect the integrity of the whole structure?

PROVISION FOR MOVEMENT

A building free of external restraint is free to move under expansion forces. Where one building abuts another it is less free to move against its neighbour unless a movement joint is incorporated. (See figure 9.)

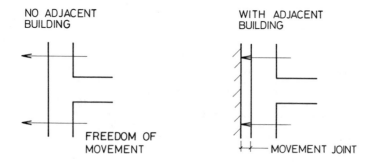

Figure 9. Horizontal freedom for movement.

PROVISION FOR FLEXIBILITY IN COLUMNS AND STANCHIONS

Columns and stanchions should be so disposed that they can accept the greatest movement across their weaker axis. (See figure 10.)

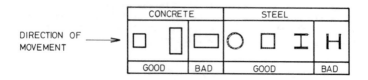

Figure 10. Flexibility in movement.

INTEGRITY OF COMPARTMENTS

Walls and floors serving as compartments should be detailed to ensure no break in integrity whilst allowing flexibility of movement. (See figure 11.)

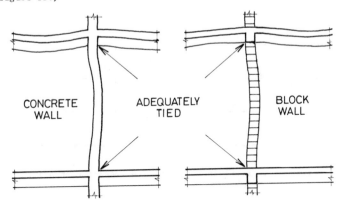

CONCRETE WALL

ADEQUATELY TIED

BLOCK WALL

Figure 11. Integrity of compartments.

LOAD RE-DISTRIBUTION

Slabs and beams so detailed by the use of suitably disposed reinforcement can re-distribute load by arching action to the supports or by the development of suspension systems from within. (See figure 12.)

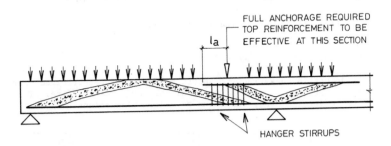

FULL ANCHORAGE REQUIRED
TOP REINFORCEMENT TO BE
EFFECTIVE AT THIS SECTION

l_a

HANGER STIRRUPS

Figure 12. Load re-distribution.

MAINTAINING STABILITY

The collapse of one structural member can bring about local instability or loss of overall instability. (See figures 13 and 14.)

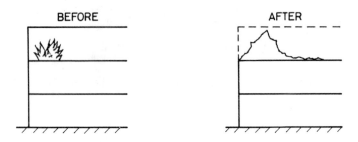

Figure 13. Local instability from loss of roof beam.

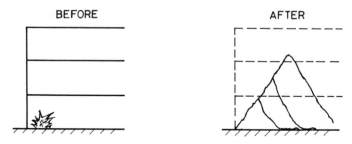

Figure 14. Overall instability from loss of beam/slab.

PERFORMANCE OF JOINTS

All joints in the structure should be checked for ability to carry load across the joint whilst maintaining integrity against fire. There should be no straight path for flame passage. (See figure 15.)

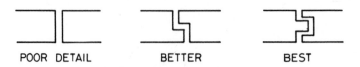

POOR DETAIL   BETTER   BEST

Figure 15. Joint configurations to resist fire.

125

STRUCTURAL PENETRATION BY SERVICES

Services to buildings viz water, gas, electricity, communication
cabling, document conveyors and the like have to penetrate structure in
practical designs. Where possible such services should be run in
protected shafts horizontally and vertically. (See figure 16.)

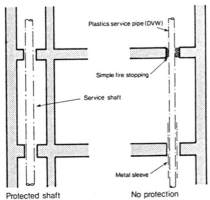

Figure 16. Provision for Services.

CONCLUSION

The design of the whole structure against fire entails more than
just compliance with building regulations for individual elements of the
structure. An understanding of the expected behaviour of the entire
structure exposed to fire is needed to ensure that the objectives of
safety of life and property are met.

ACKNOWLEDGEMENTS

The author acknowledges, with thanks, Mr. H.L. Malhotra for his
permission to reproduce extracts from his paper to the FIP Congress in
Stockholm in 1982 "Experience of Recent Fires in Concrete Structures" and
also the Institution of Structural Engineers for their permission to
reproduce extracts from their publication "Fire Resistance of Concrete
Structures".

DEVELOPING DESIGN CONCEPTS FOR STRUCTURAL FIRE

ENDURANCE USING COMPUTER MODELS

David C. Jeanes, P. E.

Director - Codes and Standards
American Iron and Steel Institute
Washington, D. C. USA

The development of appropriate computer models to predict the response of buildings to fire is now making it possible to analyze specific structures under realistic fire exposure conditions. Unlike traditional test methods and empirically derived calculation solutions, the computer-generated approach provides the engineer with a solution considerably closer to reality. Research and development work continues in making better utilization of the computer generated solutions with the aim of eventually providing the design engineer with a solution technique based on sound scientific and engineering practices.

Work sponsored by many research and industry organizations has resulted in the development of analytical methods to predict fire growth, heat transfer, and structural response.

An evaluation of the published work is now underway to develop the appropriate methodology. As part of this process, several computer models have been found to reliably predict conditions of structural fire endurance and are now being put to practical application.

INTRODUCTION

The analysis of structural fire resistance is a complicated process because of the many variables involved. These variables include fire growth and duration, temperature distribution in the structural elements, interaction between the building components, changes in material properties, and the influence of loads on the structural system. For this reason the building codes and regulation in the U.S. have relied on standardized test methods (1) to specify fire endurance requirements. Fire endurance times are assigned by the building codes for various portions of the assembly, depending on its relative significance to the overall structural stability. The primary objective of the test method is to determine the length of time that a structural assembly will withstand exposure to the test conditions. While this approach provides a reasonably simple solution to an otherwise complex problem, it does not provide the designer with a prediction of actual structural performance.

Until recently the designer has not played a part in assessing structural fire endurance requirements. The structural design would be made largely independant of any consideration of the thermal effects of the fire. Fire protection would then be added on to the completed assembly in acccordance with the established test ratings. With the cost of fire proofing typically representing as much as 20% of the cost of a structural frame and with attempts to better define structural conditions in the test furnace, the engineer is becoming increasingly more concerned with the proper design for fire endurance.

A realistic fire endurance analysis can be made based on established engineering principles. Using computer models, this approach is becoming increasingly more practical. With the development of this technology the designer is better able to evaluate the influence of structural response on the performance of supported utilities and systems, the effect of compartment size and "real" fires on exposure severity, and the potential damage to unexposed portions of a building.

Structural Fire Endurance

The structural fire endurance of a building system is a measure of its ability to resist collapse during exposure to a fire. The approaches available to make this assessment range from the use of standardized laboratory tests to the application of engineering methods. In either case, a certain level of damage is acceptable provided it does not result in the collapse of any part of the assembly or contribute to the spread of fire, smoke or hot gasses. The influence of a building fire on a structural steel frame is not generally significant until or unless the fire becomes fully developed. Therefore, the initial period of fire growth is not usually considered as part of the structural fire endurance time period.

The standard test method, ASTM E119, allows comparisons to be made of different types of construction, but provides limited information on actual performance of a building during exposure to fire, Figure 1. The criteria for acceptable performance in the case of laboratory tests includes visual observations along with the measurement of temperatures in the structural elements. The critical temperature limits are defined by the relationship between the reduction in material strength and usual design stress levels.

The test assembly is not continuous beyond the furnace walls and does not consider the effect of the fire on the unexposed portion of the structure. However, because the interaction of the framing system is known to influence structural performance, the test method does attempt to recognize the effect of the surrounding structure on resisting expansion of the assembly.

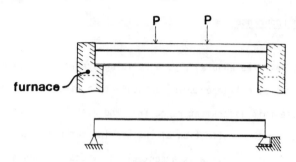

FIGURE 1. Typical Support Conditions of Floor/Beam Assembly in Standard Furnace Test (ASTM E 119).

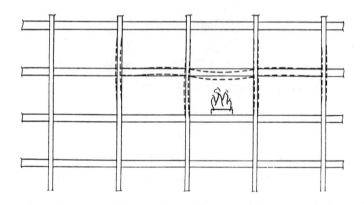

FIGURE 2. Representative Response of Structural Building Frame to Fire Exposure.

In contrast to the limited evaluation of an isolated assembly represented by the test method, the manner in which a real structure performs during a fire is influenced to varying degrees by the interaction of the connected structural elements in both the exposed and unexposed portions of the assembly, Figure 2. In general, the structure experiences both axial and moment restraint along the boundaries of the fire exposed area. Axial restraint is difficult to quantify because of constantly changing conditions. It is influenced by both the relative stiffness of the structure and localized details. These restraining conditions occur at the boundary and supports of the exposed assembly. The result is that the surrounding frame tends to resist further deflection of the exposed portion. In general, the greater the deflection the more significant is the restraining action, both axial and moment, on overall fire endurance.

The analysis of the conditions described above, superimposed on the structural assembly, can be accomplished using established principles of engineering mechanics. The analysis, however, needs to recognize the continuingly changing properties of the materials at elevated temperatures and inelastic deformation characteristics. With the ability of the computer to quickly solve iterative type problems, modeling techniques have been developed making it possible to provide this kind of analysis for steel framed floor systems.

DEVELOPING A METHODOLOGY

The development of a method for the rational determination of structural
fire endurance is a complex process.To evaluate all aspects of the problem the
solution needs to consider three distinct components: the fire exposure, the
transfer of heat from the fire to the structure, and the response of the
structure. The solution is complicated by the many variables affecting each
of these components and by the expertise necessary to assess each one.
Combustion chemistry tells us the way fires grow. Thermodynamics explains how
heat is transferred from the fire to the structure. Metallurgy defines the
effects of high temperatures on the properties of the structural steel.
Statistical methods help identify the probable risk. The building authorities
specify the level of acceptable performance. A proper design method needs to
account for the combined effect of all the prescribed conditions.

In order to develop an engineering methodology, a program has been
initiated at Worcester Polytechnic Institute (Worcester, Massachusetts) (2).
The initial objective of this program is to establish a systematic approach
which defines the interrelationship associated with each of the three
components of the fire problem and to identify the many design parameters.
With this done, a survey was initiated to examine the state-of-the-art
technology now becoming available to the designer. Computer models which
represent the most significant work, have been identified Figure 3. Those
models selected to address various aspects of the solution are now being
evaluated to verify their acceptability. In general, it presently appears

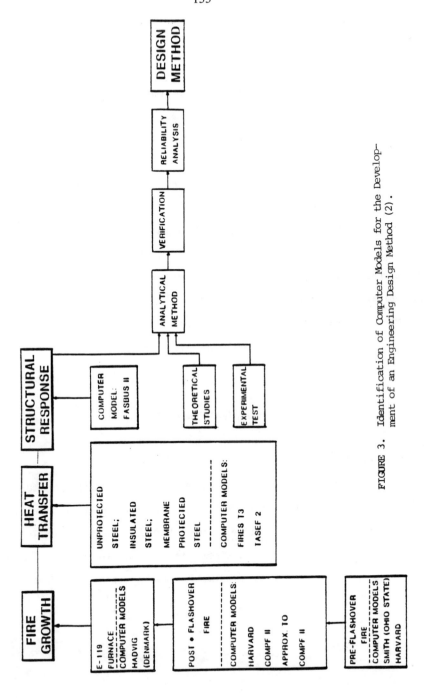

FIGURE 3. Identification of Computer Models for the Development of an Engineering Design Method (2).

134

that the solution to each component of the problem, fire growth, heat transfer and structural response, can be reliably solved independently of each other.

As this process continues certain of the computer models will be used to conduct sensitivity analyses. From these analyses, key design parameters can be identified. Statistical methods will then be applied to evaluate the probability of certain exposure conditions developing and the influence of the these conditions in combination with other loads on the structure. When these studies are completed, it is anticipated that a significant simplification in the analysis will be realized resulting in a design method for fire endurance integrated as part of the basic engineering calculations for steel structures.

Several computer models under study as part of this developing methodology provide for the evaluation of heat transfer and structural response. The use and application of two models as "tools" for evaluating structural fire endurance of steel framed floor systems are discussed in the following sections.

HEAT TRANSFER MODEL

The ability of a building to remain stable during exposure to a fire has for a long time been equated to temperature rise in the exposed structural elements. This approach is based on the fact that the mechanical properties of structural materials are reduced as the temperature of the material increased. The changes in material properties most significant to structural performance are: yield strength, modulus of elasticity and coefficient of thermal expansion. The critical level is generally defined as the temperature at which the yield strength of the material is reduced until it nearly equals the design stress and therefore, the "factor of safety" approaches unity. However, using temperatures as input, a structural analysis can be made to more accurately predict performance.

Numerous fire endurance tests have been run over past years for various size members, types of fire protection materials and thicknesses of application. From this data base, certain systems and materials have demonstrated consistantly reliable performance. By characterizing the properties of these "proven" materials, numerical techniques for solving the heat transfer problem can be approached with reasonable confidence.

The factors influencing the heating of a structural member include: the thermal properties of the materials, the mass of the structural element, the surface area exposed to the fire, and the intensity and duration of the fire. Each of these factors, inherently present in a fire test, must be specifically defined in modeling the heat transfer.

FIRES-T3 Model

The computer model (3), FIRES-T3 (FIre REsponse of Structures - Thermal - 3 Dimensional Version) is a three dimensional finite element heat transfer program. It is suitable for use in evaluating the temperature history of solid composite materials such as fire protected structural steel and reinforced concrete. A limitation of the present version, however, is that it cannot model heat transfer through cavities in an assembly.

The model allows for consideration of the nonlinear characteristics of the thermal properties of the materials and the heat transfer from the fire environment. The solution technique requires an iterative intergration process within each time step throughout the exposure period. Accordingly, the program user must exercise judgement as to the appropriateness of the solution as the analysis progresses.

The principal factors influencing its effective use are the layout of the finite element mesh and the selection of the time-step size. Both factors need to be dimensioned so that sufficient detail is available in the region and over the time period for which the thermal analysis can be expected to be most sensitive.

From a user point of view, the FIRES-T3 model allows for consideration of the following design parameters:

1. Material Properties - the thermal properties (thermal conductivity and specific heat) and density of materials are considered with respect to their change in value at elevated temperatures. (Effects of internal heat generation can also be considered).

2. Fire Environment - the time-temperature history of the fire environment is considered by specifically defining the temperature at each time step during the solution. Therefore, the fire exposure curve can take any form (ie. constant temperature, linear change, E119 curve or natural burning).

3. Heat Transfer - the heat transfer process due to the fire exposure is modeled as convection and radiation in the fire boundary and as conduction through the member. The emissivity of the flame and surface, view factor, and surface absorption are considered in

calculating radiation effects. Convection is modeled using a convection factor and power of convection. Conductivity is computed using the appropriate material properties.

4. Geometry - the shape and size of the structural element can be considered in one-, two-, or three dimensions. This is accomplished by drawing a mesh representing the shape and arrangement of materials of the element and describing this arrangement in terms of the coordinates of each of the nodal points in the mesh.

The output of the FIRES-T3 analysis provides a listing of the calculated temperature at each node, the average temperature of each element in the mesh, and a summary of the test conditions at that point in the analysis.

In order to evaluate the accuracy of the FIRES-T3 program, the program was first used to model assemblies for which actual test data was available. This approach allowed for confidence to be established in the model without a need to be able to specifically understand all modelling techniques used.

Predictions were made of the heat transfer through steel beams with direct-applied fire protection material from assemblies tested at Ohio State University, Underwriters Laboratories and the U.S. National Bureau of Standards. The modeling was done using the nodal mesh illustrated in Figure 4. The results demonstrated good agreement between the predicted and recorded average section temperatures and the temperature profiles through the sections, Figure 5.

As a result of the satisfactory agreement demonstrated by this modelling, a series of analysis were conducted in order to develop data useful as design

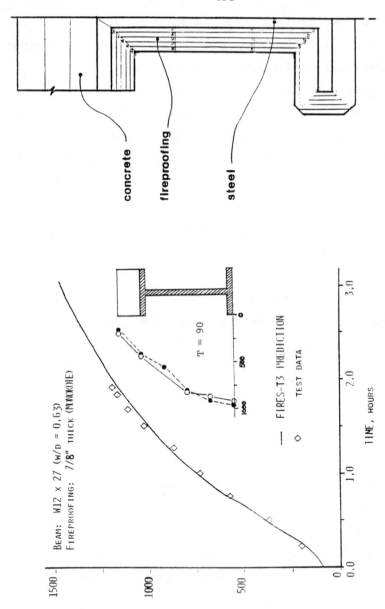

FIGURE 4. Finite Element Mesh Used for FIRES-T3 Analysis of Steel Beams with Direct Applied Fire Protection.

FIGURE 5. Comparison of Temperatures Predicted Using FIRES-T3 and Corresponding Test Data.

aids. This was done by analyzing different size steel beams with direct
applied fire protection thicknesses of 1/2, 1, and 1 1/2 inches. The beams
were selected to cover a range in W/D values from 1.5 to 2.5. The fire
exposure used in the analysis was the ASTM E119 time-temperature curve over a
four hour period. The results of this series of analysis have been compiled
and presented as "Fire Endurance Time versus W/D", Figure 6. This general
form of the data utilizes the W/D characteristic of the beam as the basic
design parameter. The data presented is based on the average section
temperature of 1000F (538C) for the criteria.

STRUCTURAL RESPONSE MODEL

The structural fire endurance of a building system is a measure of its
ability to resist collapse during exposure to a fire. The approaches used to
make this assessment range from the use of standardized laboratory tests to
the application of engineering methods. In either method a certain level of
damage is acceptable provided it does not result in the collapse of any part
of the assembly or contribute to the spread of the fire. The influence of a
building fire on the structural steel frame is generally not significant until
or unless the fire becomes fully developed. Therefore, the initial period of
fire growth is not usually considered as part of the structural fire endurance
time period.

The analysis of a structure exposed to fire can be accomplished using the
established principles of engineering mechanics. The analysis, however, needs
to recognize the continuingly changing properties of the materials at elevated
temperatures. Those properties which are most significant to structural
performance are: yield strength, modulus of elasticity, and coefficient of

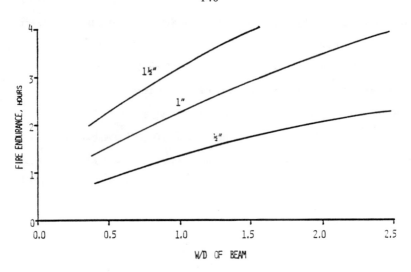

FIGURE 6. Fire Protection Thickness* for Steel Beams Based
on Average Section Temperature Limit of 1000F
(*cementicious and mineral fiber direct applied).

a. Beam Element

b. Slab Element

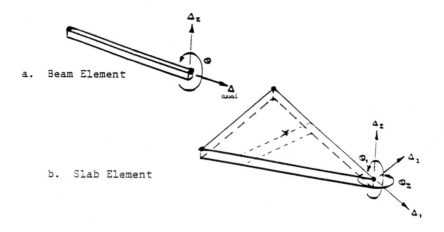

FIGURE 7. Finite Elements Used in the FASBUS II Computer Model.

thermal expansion. Studies have been made to characterize the changes in these properties with temperature. Utilizing this data, structural fire endurance can be determined by repetitive calculations. Because of the ability of the computer to quickly solve these tedious types of problems, modelling techniques have been developed making it possible to provide this kind of analysis for steel framed floor systems.

FASBUS II Computer Model

The computer model, FASBUS II (FIRE Analysis of Steel BUilding Systems) is a structural analysis program specifically designed to analyze the fire endurance of steel framed floor systems.   The model utilizes the finite element method where beam elements and triangular plate bending elements are used to represent the frame and slab, respectively, Figure 7.

The incremental solution used by the model provides for consideration of changes in temperature, with corresponding changes in material properties, throughout the exposure period. Using an iterative process the model determines the displacements necessary to bring the structure to a point of static equalibrium under the loads and conditions imposed.

As with any engineering analysis, the designer must have a basic understanding of the problem being solved and the solution techniques applied. Accordingly, the user of the FASBUS II computer program should have a basic knowledge of structural mechanics, an understanding of the modeling techniques and a familiarity with of building construction and the thermal effects of a fire. With this background the user will be able to more accurately define the physical characteristics of the problem and express them in terms identifiable to the computer model.

From a user point of view, the model provides for consideration of the following design parameters:

1. **Geometry of Structural Elements** - In addition to the layout of the framing members, detailed description of the structural elements is permitted which includes shape and placement of the steel beam sections, deck profile and reinforcement locations.

2. **Material Properties** - Non-linear changes in the yield strength, modulos of elasticity, and coefficient of thermal expansion with respect to temperature are inputted directly (material models within the program allow for consideration of the elastic/plastic character of steel and cracking or crushing of concrete).

3. **Loads and Restraint** - Provision is made for the direct input of point loads (any direction) and uniform vertical loads acting on the floor system. The resistance of structural elements connected to the assembly, such as columns and braces, can be modeled.

4. **Time-Temperature Exposure** - The shape of the temperature profile with respect to time of exposure for up to five groups of elements in the model can be specified. Such profiles are based on either measured or calculated data which reflect the nature of the fire exposure being considered.

The results of a FASBUS II analysis provide the designer with predictions of deflections and rotations across the floor system and stress and strain conditions within the structural members.

Evaluation of the Model

The computer model is based on the mathematical relationships which best define the engineering mechanics applicable to the structural system under consideration. As with any such approach, certain assumptions need to be made to provide a model which is suitably efficient for both design and analysis applications. It is, therefore, desirable to be able to verify the theory and the assumption contained in the model through the demonstrated agreement between actual and predicted responses of fire exposed assemblies.

During the development of the FASBUS computer model, it was routinely used to evaluate the results of various fire endurance tests [3,4]. The assemblies investigated included floor slabs with and without framing members, beams with composite and non-composite slabs and a reinforced concrete slab model. In all cases, the assemblies were exposed to the time-temperature curve defined by ASTM E119. In general, it was found that the measured performance was accurately predicted where sufficient detail was recorded to clearly describe the test assembly (with boundary and support conditions) and the structural performance during the fire exposure.

Because the engineering approach represented by the computer model is a significant departure from the laboratory test methods referenced in the building codes, a substantial evaluation of the computer model was necessary. The analysis of data collected from a large scale test program conducted at the U. S. National Bureau of Standards provided the basis for this evaluation (6). The test program provided for the measurement of the response of a structural system representative of actual building construction, when a portion of that assembly is exposed to fire. This was accomplished by recording vertical and horizontal deflections of the frame and floor slab and temperatures of the exposed structural components. The actual test assembly consisted of a two story-four bay structural steel frame with a concrete and steel deck floor slab, Figure 8. A total of three tests were conducted on the assembly which included both controlled.exposure fires (ASTM E119) and a free burning "real" fire.

Evaluations of the fire exposure conditions recorded during each of the tests have been made using the finite element mesh illustrated in Figure 9. Comparisons between the recorded and predicted performance for each test, demonstrated good agreement for both the deflected shape of the floor assembly and level of damage to the concrete slab and steel frame. These comparisons for a 90 minute exposure to the ASTM E119 time-temperature curve are illustrated in Figure 10.

145

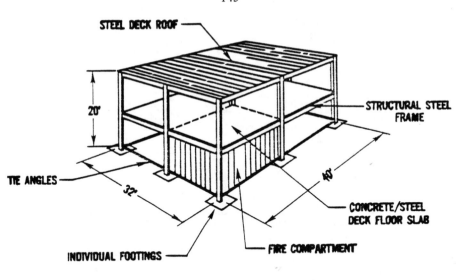

FIGURE 8.   Structural Fire Endurance Test Frame at the U. S.
National Bureau of Standards (NBS).

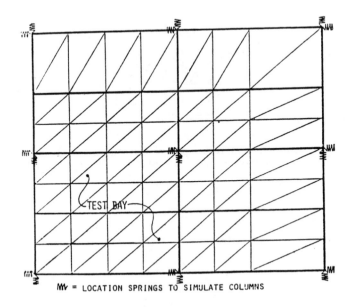

MW = LOCATION SPRINGS TO SIMULATE COLUMNS

FIGURE 9.   Finite Element Mesh Used to Model NBS Test Frame.

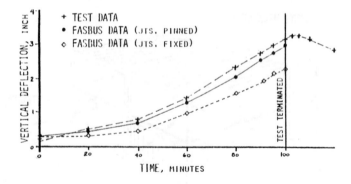

a. Vertical Deflection at the Center of the Test Bay During the Course of the Test.

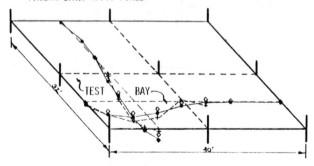

b. Vertical Deflection Across Center of Test Bay Floor Slab (@ 90 min., E 119).

FIGURE 10. Comparison of FASBUS II Predictions with Test Data.

147

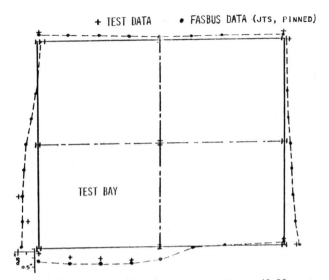

c. Lateral Deflection of Test Floor Frame (@ 90 Min., E 119).

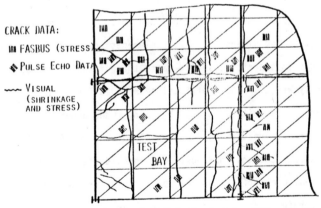

d. Comparison of Recorded Slab Crack Data (After Test).

FIGURE 10 (Contd.)

APPLICATION OF THE MODELS

Once the computer models have been developed and adequately validated they can be considered as engineering "tools" in the design of structural elements. The computer models, FIRES-T3 and FASBUS II, are more specifically analytical programs than design programs. As such, the models are used to evaluate a certain set of exposure conditions of a particular structural assembly. Using the results of the analysis the designer can then evaluate the acceptability of a given design.

The scope of the analysis must first be determined. This includes identifying the assembly or portion thereof to be modeled, the temperature conditions of the exposure, the distribution of live loads (or load combinations) and the types of materials and construction represented. The structural assembly to be analyzed must then be redefined in the form of the element types included in the model. The size of the elements is determined by the dimensions of the "nodal mesh" into which the assembly is divided. The spacial position and dimensions of the elements, defined by the nodal coordinate system therefore, should match that of the actual assembly.

The application of the analytical techniques for determining structural fire endurance is permitted in U.S. building codes under the general provisions for "alternates" to the prescribed code requirements. To exercise these provisions, the designer must produce evidence sufficient to satisfy the responsible building official. The specific requirements for a particular case will therefore vary according to the level of interest and expertise of the individual official reviewing the analysis.

Typical of the acceptance of most new design approaches, is the need for a project significant enough to warrant the interest of the designer. The following is a brief description of a successful analysis, using the two computer models described above, which resulted in establishing a change to the fire protection requirements on portions of a structural frame.

Evaluations of An Office Building

The first application of the analytical approach represented by the FIRES-T3 and FASBUS II computer models was made on a 42 story office building located on the West coast of the U.S. (7). The interest of the designer developed when he considered the excessive fireproofing requirements placed on the large spandrel beams of the building frame. The specific beams, which were designed to carry earthquake forces, provided substantial bracing to the structure through moment connections to the columns. Despite the structural conditions, fire endurance requirements specified were initially based on a standard fire test rating of a simply supported beam of a size significantly smaller than those in the structure. In addition, the code required a three hour rating for these beams, because of their function in stabilizing the columns.

The building designer was only interested in evaluating the spandrel beams, thereby limiting the analysis to the assembly shown in Figure 11. Using a direct applied fireproofing thickness of 3/4 inch (1.9 cm) a FIRES-T3 analysis was conducted on each of the spandrel beams of size W33x118 and larger. The results of the analysis after 1-1/2 hours and 3 hours of exposure for the W33x118 beam predicted a high point temperature of 1490F (810C) and an average section temperature of 1300F (715C), Figure 12.

Using the predicted temperature history, a structural analysis was made using FASBUS II. The modeling considered only the gravity loads supported by the beam thereby ignoring the higher stress levels due to earthquake loading. The influence of columns was modeled as equivalent stiffnesses applied to the

150

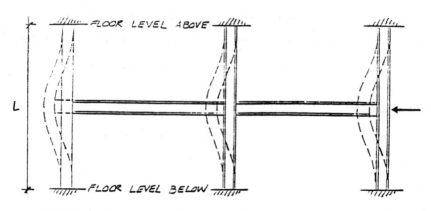

FIGURE 11. Portion of 42-Story Office Building Analyzed Using
FIRES-T3 and FASBUS II.

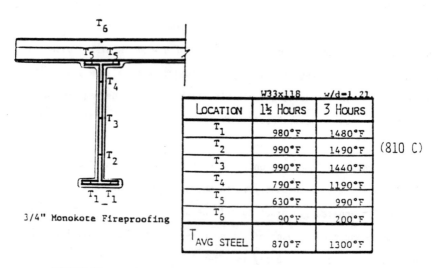

3/4" Monokote Fireproofing

LOCATION	1½ HOURS	3 HOURS
T_1	980°F	1480°F
T_2	990°F	1490°F
T_3	990°F	1440°F
T_4	790°F	1190°F
T_5	630°F	990°F
T_6	90°F	200°F
$T_{AVG\ STEEL}$	870°F	1300°F

W33x118   w/d=1.21

(810 C)

FIGURE 12. Temperature Predictions in W 33 x 118 Spandrel
Beam Using FIRES-T3.

ends of the beam. From the results of the analysis the vertical deflection and elongation of the beam could be examined over the course of the exposure period, Figure 13. In addition, the stress levels across the beam section were evaluated to determine the development of plastic material conditions.

The combination of the heat transfer analysis and the structural response modeling was satisfactory to the building official. As a result the thickness of the fire protection material on all the spandrel beams equal to or larger than the beam analyzed was reduced to 3/4 inch (1.9 cm).

SUMMARY

The development of an engineering method for calculating structural fire endurance of steel buildings is now under development in the U.S. Based on a study of these design parameters, a systematic approach has been defined which identifies the various components of the design problem and their interrelationship. Examination of the state-of-the-art technology available for addressing each of the design parameters is now being accomplished. Because of the complexity of the problem computer models are commonly required.

Two computer models, FIRES-T3 and FASBUS II, have been developed to predict heat transfer and structural response, respectively. As a result of a substantial evaluation of each of these models, their validity to accurately predict structural fire endurance of typical steel framed floor constructions has been established. The application of these models to actual building constructions has demonstrated their value as an engineering "tool".

152

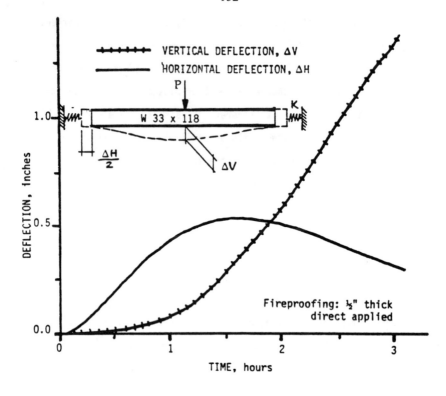

FIGURE 13. Predicted Performance of a Fire Exposed Steel Spandrel
Beam with Applied Loads and Restraining Conditions
(using FASBUS II).

153

With the continued development of an engineered solution, these and other
computer models are being used to identify and evaluate each of the
parameters. The most significant of these parameters will be used to
developed a concise and optimum design. It is anticipated that the design of
structures to resist the effects of a building fire will eventually become a
routine part of the structural design of the building frame.

REEFERENCES

1) "Standard Methods of Fire Tests of Building Constructions and
   Materials", ASTM E119, Annual Book of Standards, Part 18, pp.
   941-967, American Society for Testing and Materials, 1982.

2) Fitzgerald, R.W., Development of an Engineering Method to Calculate
   the Fire Resistance of Structural Steel Frames, Status Report to
   American Iron and Steel Institute, Worcester Polytechnic Institute,
   December 1982.

3) Iding, R.J., Bresler, B. and Nizamuddin, Z., "FIRES-T3, a Computer
   Program for the Fire Response of Structures-Thermal", Report No. UCB
   FRG77-15, University of California, Berkeley, October 1977.

4) Chiapetta, R.L. et al, "The Effect of Fire Temperatures on Buildings
   with Steel Frames", Final Report IITRI Project J8095, Chicago, ILL.,
   April 1972.

5) Iding, R.H. and Bresler, B., "Effect of Fire Exposure on Steel
   Framed Buildings", Report to American Iron and Steel Institute, WJE
   No. 78124, Wiss, Janney, Elstner and Associates, Inc., Emeryville,
   CA., March 1982,

6) Jeanes, David C., "Predicting Fire Endurance of Steel Structures",
   Preprint 82-033, American Society of Civil Engineers, ASCE
   Convention, Nevada, April 16-30, 1982.

7) Bresler, B., Iding, R., Amin, J., and Laws J., "Evaluation of
   Fire Proofing Requirements for a High-Rise Steel Building" paper
   presented at the AISC National Engineering Conference, Memphis, TN.,
   April 1983.

CALCULATION METHODS FOR FIRE ENGINEERING DESIGN OF STEEL AND COMPOSITE
STRUCTURES.

L. TWILT and J. WITTEVEEN

Institute TNO for Building Materials and Building Structures
Rijswijk, THE NETHERLANDS

INTRODUCTION

The strength and deformation properties of structural steel are
fairly rapidly impaired with rising temperature of the steel.
Consequently, a steel or a composite structure, when exposed to fire, is
endangered, since the structure will deform or even collapse. For reasons of safety or – more likely – loss prevention, it may be justified
to require that the structure withstands the anticipated fire for a
certain period of time. This calls for a proper structural fire engineering design.

The available methods for structural fire engineering design can be
categorized as follows [1], [2]:

– Level (1): Method on the basis of ISO standard fire exposure.
The design criterion is that the fire resistance is equal to or exceeds the time of fire duration required by building regulations or
codes.
– Level (2): Method on the basis of a standard fire exposure.
The design criterion is that the fire resistance is equal to or exceeds the design equivalent time of fire exposure, a quantity which
relates compartment (non-standard) fire exposure to the ISO standard
fire.
– Level (3): Method characterized by a direct analytical design on the
basis of compartment (non-standard) fire exposure.

A design procedure according to level (1) corresponds to a vast
majority of national building codes, in which the requirements are
expressed as a required time of fire duration, i.e. the fire resistance
time, directly related to the standard fire. In most countries the level
(2) and level (3) methods have occasionally been used but, except in
Sweden, they are not yet automatically accepted as methods which satisfy
the requirements of the building regulations.

Until recently, the structural fire engineering design could only
be based on fire resistance tests. Such a procedure is time consuming
and expensive, and sometimes gives rise to anomalies due to variation in
test results. During the last ten years, however, important progress has
been made in the development of analytical models, by which the fire
resistance of structural building components can be determined. The analytical methods lead to more defined and uniform levels of safety and
– under circumstances – to more simple and systematic design procedures,

156

and must therefore be considered as a more appropriate tool for a fire
engineering design.

Within the European Convention for Constructional Steelwork (ECCS),
Technical Committee 3 "Fire Safety of Steel Structures", has elaborated
calculation methods for the fire resistance of both traditional structural steelwork and composite elements [3], [4], [5]. Aim of this work
is to provide a reference for national codes of practice with the final
objective of achieving a single European code.

This paper reveals the basic features, possibilities and limitations of the above mentioned ECCS-calculation methods.

SCOPE AND BASIC FEATURES OF THE ECCS-CALCULATION METHODS

The ECCS-calculation methods are developed to provide an alternative for the standard fire resistance test [6]. They concentrate therefore on a level (1) fire engineering design as identified in the introduction. It is noted, however, that with some restrictions, they can
also be used as a component in a level (2) design. A level (3) design
(i.e. based on the natural fire concept) is not covered by the
calculation methods. The analytical design concept on a level (1) basis,
is illustrated in Fig. 1. Essentially, it should be considered as a
verification procedure.

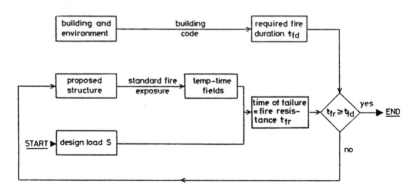

Fig. 1: Analytical fire engineering design based on thermal exposure
according to the standard fire resistance test.

The design criterion itself (i.e. the required fire resistance) is
not dealt with in the ECCS-calculation methods. This is left to the
national authorities. According to their aim of providing an alternative
to the standard fire resistance test, the ECCS-calculation methods only
give rules how to calculate the fire resistance of building components.
The following possibilities, corresponding to common practice in fire
testing, are presented:
  - statical determinated beams or slabs
  - statical indeterminated beams or slabs
  - centrically loaded columns.

In the case of composite structures the analysis is limited to composite concrete slabs with profiled steel sheet and to composite columns.
The necessary calculations include to main steps:

- A calculation of the temperature distribution within the fire exposed structural element or assembly during the heating process.
- A transformation of these temperature distributions to the variation of the load bearing capacity as a function of time in order to examine whether or not the fire exposure will cause a failure of the structural element or assembly at the specified loading.

The combination of these two steps gives the time that the structure can withstand the fire. For a level (1), (2) approach this period of time corresponds to the fire resistance of the structural element or assembly.

Calculating the temperature distribution in the building element for tradiotinal steelwork a one dimensional heat flow is assumed. For composite members, in general the use of two dimensional heat flow models is necessary. The effect of the fire characteristics and the geometry of the structural element as well as its protection – if any – is accounted for.
The starting point in determining the load bearing capacity is a proper description of the mechanical properties at elevated temperatures of the structural materials involved. The calculation itself is based on the theory of plasticity.

Where appropriate, the ECCS-calculation methods are formulated in such a way that conservative but simple solutions are obtained. A basic feature in this respect is the assumption that the mechanical properties of steel and concrete are independent of time, i.e. creep effects are included implicitly. As a direct consequence, the above mentioned calculation steps may be dealt with seperately. Thus, a significant simplification of the analysis is achieved. In those cases, where the effect of deformations on the load bearing capacity may not be ignored (e.g. columns), approximate rules, closely related to the calculation procedure at room temperature are given, as an alternative to more exact but complicated solutions.

TRADITIONAL STEELWORK

The calculation method for traditional steelwork is given in the European Recommendations for the Fire Safety of Steel Structures [3] and is based on the extensive information, experience and studies from many countries. A proper validation of the given rules was therefore feasible [7]. Reason why the calculation method has been presented as formal Recommendations. In the following discussion some of the main aspects of these Recommendations will be evaluated.

Thermal Response

Generally, it is necessary to calculate transient heat conduction to the steel section. Where an analytical solution is not available, it is possible to adopt a quasi-stationary approach, iterated for successive time intervals. This calculation method can be simplified by considering the steel to be a heat sink, with negligible resistance to heat flow; thus any heat supplied to the steel section is considered to be instantly distributed to give a uniform steel temperature.

Under these conditions the temperature distribution in the steel can be calculated with classical one-dimensional heat flow theory [8] [9], [10], [11]. The Recommendations distinguish between:

- unprotected steel members;
- lightly insulated steel members;
- heavily insulated steel members.

Under the given assumptions, the resistance of unprotected steel members to heat flow is governed only by convection and radiation. The coefficient of heat transfer due to convection from the fire to the exposed surface of the steel member, α_c, is considered to be constant with a value: $\alpha_c = 25$ W/m^2·°C. The coefficient of heat transfer due to radiation, α_r, is a function of the gas and steel temperatures and can be determined from the Stephan Boltzman law of radiation. The resultant emissivity ε_r of the flames, gases and exposed surfaces which appears in this formula, is considered constant with a value of $\varepsilon_r = 0.5$, giving a conservative solution. The Recommendations give tabulated values from which the temperature development of unprotected steel members under standard fire exposure can be obtained as a function of the "section factor" F/V. In Fig. 2 a graphical display of such a temperature-time relationship is given together with the relevant formula.

The resistance to heat flow of insulated steel members is governed by convection, radiation and the thermal conductivity of the insulation material. For practical applications however, the influence of convection and radiation can often be neglected. Additionally, a distinction is made between lightly insulated members, for which the heat capacity of the insulation material can be neglected, and heavily insulated members, for which the heat capacity of the insulation is taken into account.

The Recommendations give tabulated values from which the temperature development of protected steel members under standard fire exposure can be obtained. The values are presented as a function of the "section factor" F/V and the insulation characteristic, d/λ, in which d is the thickness of the insulation and λ the thermal conductivity of the insulation.

Fig. 3 exemplifies graphically such a temperature-time relationship for a lightly protected steel member together with the relevant formula.

It is emphasized that the value of the thermal conductivity of the insulation material, λ, is generally not identical to the conventional value as given in the handbooks on heat transfer. The value of λ will depend on the temperature as well as on the deformation capacity of the material attached to the steel member. During deformation, cracks or openings may occur. In order to include these effects, apart from small scale experiments for determining the thermal conductivity of the insulation material, at least one full scale test on a loaded member must be performed.

159

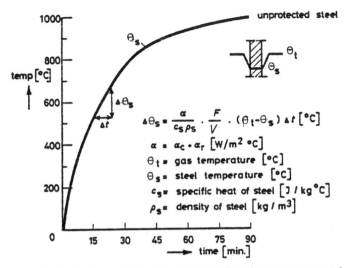

$$\Delta\Theta_s = \frac{\alpha}{c_s \rho_s} \cdot \frac{F}{V} \cdot (\Theta_t - \Theta_s) \Delta t \; [°C]$$

$\alpha = \alpha_c + \alpha_r \; [W/m^2 \; °C]$

$\Theta_t =$ gas temperature $[°C]$

$\Theta_s =$ steel temperature $[°C]$

$c_s =$ specific heat of steel $[J / kg \, °C]$

$\rho_s =$ density of steel $[kg / m^3]$

Fig. 2: Graphical display of tabulated values for the temperature-time development in an unprotected steel member (F/A = 100 m^{-1}).

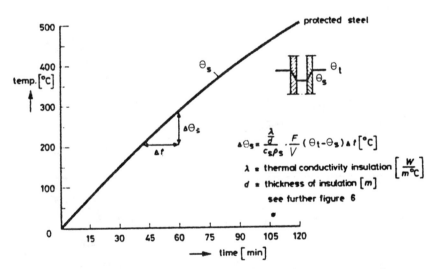

$$\Delta\Theta_s = \frac{\frac{\lambda}{d}}{c_s \rho_s} \cdot \frac{F}{V} (\Theta_t - \Theta_s) \Delta t \; [°C]$$

$\lambda =$ thermal conductivity insulation $\left[\dfrac{W}{m \, °C}\right]$

$d =$ thickness of insulation $[m]$

see further figure 6

Fig. 3: Graphical display of tabulated values for the temperature-time development in a lightly protected steel member (F/A = 100 m^{-1}; d/λ =0.2 m^2 °C/W).

160

For practical applications and under certain conditions, use can be made of more approximate design equations, as an alternative for the above mentioned tabulated design data [7]. Such equations give the time, necessary to reach a certain steel temperature, as function of the section value and – in the case of protected members – of the insulation characteristic d/λ, and are derived by curve fitting from the "exact" calculation results. An example for steel members without insulation is presented in Fig. 4. Some test data are also given. There is a reasonable agreement between theory and experiment.

In a more generalized form, the equation for the unprotected members can be written as:

$$t = 0.54 \ (\theta_s - 50) \ (F/V)^{-0.6} \tag{1}$$

with θ_s = steel temperature in °C, F/V section value in m^{-1} and t = time to reach θ_s in min. Due to its curve-fitting background, the range of application of equation (1) is limited. The following limits hold:

t = 10 – 80 min.
θ_s = 400 – 600°C
F/V = 10 – 300 m^{-1}

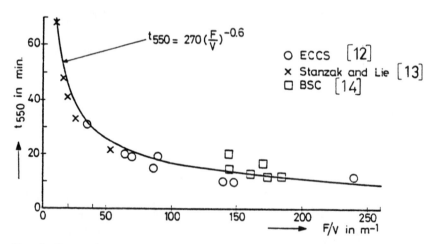

Fig. 4: Measured temperatures of members without insulation.

As exemplified in Fig. 5, similar design equations can be derived for protected members. Besides the section factor F/V, now also the insulation characteristic d/λ plays a role. Again, proper limits should be set to the range of application. It will be clear that design formulae as discussed here, provide a usefull tool when extrapolating test results. See [7], [15], [16].

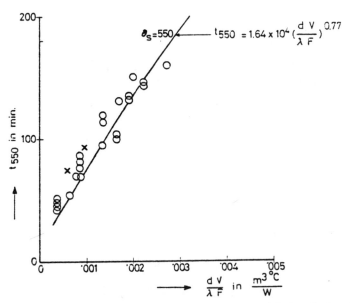

	$\dfrac{d}{\lambda}$	$\dfrac{F}{V}$
(a) O	·086 – ·259	225 – 63.1
(b) ✕	·076	133 – 82.3

(a) Vermiculite boards
(b) Sprayed light weight plaster

$$t_{550} = 1.64 \times 10^{-4} \left(\frac{d}{\lambda}\frac{V}{F}\right)^{0.77}$$

$\vartheta_s = 550$

Fig. 5: Measured times for temperature of members with insulation to reach 550°. Nominal values of λ_i assumed [7], [12].

Structural Analysis

Research reported in [17], [18] has shown that for heating rates which may be expected during fires in practice, say, between 5 and 50°C/min. and for steel temperatures not over, say, 600°C, the deformation behaviour under constant load can be considered as independent of the heating rate. Consequently a family of stress–strain relationships for different temperatures must exist, in which the influence of high temperature creep is implicitly included. In Fig. 6 the stress–strain relationships for a typical steel grade at different temperatures are presented.

162

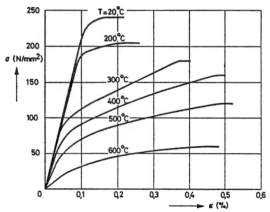

Fig. 6: Stress-strain relationships for Fe 360 at elevated temperatures.

The gap between the curves applying to 200°C and 300°C is due to so-called "thermally activated flow" [18]. Applying the elementary theory of plasticity, the curved stress-strain diagrams are cut off at certain stress levels. The horizontal plateau is defined as the effective yield stress (= $\sigma_{y\theta}$). In Fig. 7, the variation of the effective yield stress with steel temperature is given as a fraction of the yield stress at room temperature.

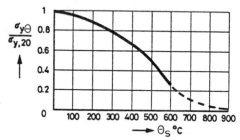

Fig. 7: Effective yield stress at elevated temperatures, $\sigma_{y\theta}$, expressed as a fraction of the yield stress at room temperature, σ_{y20} (Fe 360 - Fe 510).

Stress-strain diagrams for other steel grades up to Fe 510 (σ_{y20} = 355 N/mm²) can be obtained from the stress-strain diagram for steel grade Fe 360 by a simple transformation based on the following assumptions:
- for any grade of steel between Fe 360 and Fe 510 the effective yield stress, as a fraction of the yield stress at room temperature, is the same as for Fe 360 throughout the steel temperature range (Fig. 7).

163

- for any grade of steel between Fe 360 and Fe 510 the modulus of elasticity in the origin, $E_{o, \theta}$, is assumed to be equal to that of Fe 360 throughout the steel temperature range.

By knowing the properties of steel at elevated temperature, it is possible to calculate the mechanical behaviour of a steel structure in a fire. Due to the non-linear stress-strain relationships of steel at elevated temperatures, the linear theory of elasticity cannot be applied and use has to be made of the theory of plasticity. Two design methods are availabe, identical to those used in structural analysis at ordinary room temperature:

- a limit state design according to the elementary theory of plasticity in those cases where a similar design is allowed at room temperature;
- an incremental elasto-plastic analysis.

The first method is suitable when the limit state at elevated temperatures can be defined by structural collapse, i.e. beams in braced frames. At a given temperature the ultimate load can be calculated from the temperature dependent effective yield stress $\sigma_{y, \theta}$ (Fig. 7). This is illustrated in Fig. 8 [7] [9], [19].

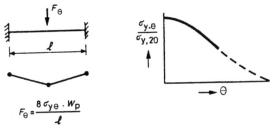

Fig. 8: Structural design at elevated temperatures according to the elementary theory of plasticity.

The second method has to be used when geometrically non-linear effects have a significant bearing on the structural behaviour i.e. columns and unbraced frames. At a given temperature, the load-bearing capacity can be determined with the appropriate stress-strain relationship (Fig.6), by computing the deflection curve [18]. Fig. 9 gives an illustration. Application of this method generally requires a computer.

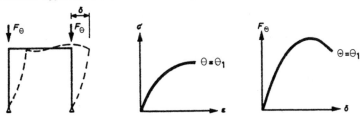

Fig. 9: Structural design at elevated temperatures with an incremental elasto-plastic analysis.

The assumption of the uniform temperature distribution in the steel member allows to define a certain unique steel temperature at which, under a given load, the steel member collapses. This temperature is denoted as the "critical steel temperature" (= θ_{cr}). When the structural analysis at elevated temperature can be based on elementary plastic theory (e.g. in the case of beams), it can be shown that, irrespective of the static system, the following simple equation holds:

$$\frac{\sigma_{y\theta}}{\sigma_{y20}} = \frac{P}{P_p} \ (= \gamma) \tag{2}$$

with:

$\sigma_{y\theta}$ = effective yield stress at elevated temperature
σ_{y20} = yield stress at room temperature
P = design load under fire conditions
P_p = collapse load at room temperature.

If the ratio $\gamma = P/P_p$ is known, the critical steel temperature can easily be determined via the relation between effective yield stress and temperature. See e.g. Fig. 7. The results can, for various values of γ be presented in a tabulated form, as exemplified in Table 1.

γ	0.3	0.4	0.5	0.6	0.7
θ_{crit} [°C]	585	540	490	430	360

Table 1: Values for the critical temperature of steel beams as a function of the degree of loading ($\gamma = P/P_p$).

When the elementary plastic theory cannot be used (e.g. in the case of columns and unbraced frames), the critical temperature should be calculated by means of a incremental elasto-plastic analysis as indicated above. For design purposes, the Recommendations give however a simplified rule for the determination of the critical temperature of centrically loaded columns. This rule is based on the following approximations [20]:

- the equation for the ECCS-buckling curves derived for room temperature conditions [21] may be used under fire conditions as well, provided that the strength and stiffness properties of steel at room temperature are replaced by temperature dependent values;
- the strength and stiffness properties of steel at elevated temperature are represented by the effective yield stress ($\sigma_{y\theta}$) and the Young-modulus in the origin (= $E_{o\theta}$) respectively;
- the ratios $\sigma_{y\theta}/\sigma_{y20}$ and $E_{o\theta}/E_{o20}$ vary in the same way with temperature.

Under the specified conditions, it can be shown that equation (2) holds for centrically loaded columns as well, provided that γ is now defined as the ratio between design load under fire conditions and the ECCS-buckling load at room temperature conditions. Accordingly, Table 1 may also be used for columns. Calculation results, based on the given rule, show good agreement with the results of full scale fire tests on columns. See Fig. 10.

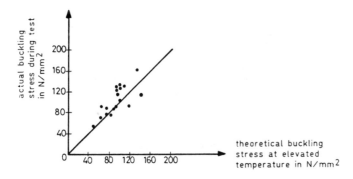

Fig. 10: Comparison of the results of columns tests at elevated temperature and theoretical predictions [20].

Consistency with Standard Fire Tests

An analytical design based upon the specified assumptions, generally results in a level of fire resistance which is conservative, in comparison with the corresponding level in a standard fire resistance test. The main reasons for this systematic discrepancy are:

- the analytical method is based upon the characteristic values of the material properties at elevated temperatures, whereas fire resistance tests are performed on specimens, the material properties of which are random samples;
- in the computations, the temperature distribution along and across the members is assumed to be uniform, whereas in tests generally a non-uniform distribution arises;
- for columns, geometrical imperfections play an important role; in tests the specimen has a random imperfection which is generally smaller than that assumed in the analytical design.

It is evident that a similar discrepancy would exist if tests at room temperature were compared with analytical results. Because most structures at room temperature are open for a complete analysis, such tests are exceptions and a discussion on the discrepancy is less relevant. However, it is likely that for the years to come, fire resistance tests as well as analytical methods are used in fire engineering design. Consequently, there is a need to develop a method for avoiding these discrepancies and getting consistency between the analytical and experimental approaches. This can be done in alternative ways.

One way is to multiply the calculated load bearing capacity – corresponding to the calculated uniformly distributed steel temperature θ_s – by a factor of magnification f, which includes corrections for representative deviations from the assumptions listed above. Such a magnification factor then can be written as:

$$f = f_m f_i f_{\theta c} f_{\theta a} \qquad\qquad (3)$$

where f_m = a correction factor related to the mechanical properties of the structural material at temperature θ_s, f_i = a correction factor related to the imperfections of the structural member, $f_{\theta c}$ = a correction factor in respect of non-uniformity in the temperature distribution over the cross section, and $f_{\theta a}$ = a correction factor in respect of non-uniformity in the temperature distribution along the structural member.

Representative values are derived in [22] for the various correction factors as well as for the total magnification factor f. Practically, f then can be given as a function of the uniformly distributed, calculated steel temperature θ_s with a differentiation with respect to columns, statically determinate beams and statically indeterminate beams with one or more redundancies.

The same results can be achieved by reducing the design load in the calculation, P, by a load multiplier $\kappa < 1$, i.e.

$$P = \kappa P_\theta \qquad\qquad (4)$$

where P_θ = the load to be applied on structural element in the fire resistance test.

Values for κ derived for beams are presented in Fig. 11.

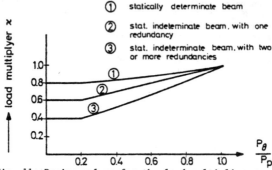

Fig. 11: Design values for the load multiplier κ.

Accordingly, a load multiplier κ should also be introduced for columns. In view of the considerable approximations necessary when deriving the calculation rule for the critical temperature of columns, it is not realistic to differentiate κ in this case. Therefore a single (conservative) value has been chosen: $\kappa = 0.85$.

COMPOSITE COLUMNS

The ECCS-calculation methods for the fire resistance of composite columns distinguish between:

(a) rolled I-profiles encased in concrete
(b) rolled I-profiles concreted between the flanges
(c) concrete filled steel sections with and without reinforcement.

In Fig. 12 these three options are reviewed.

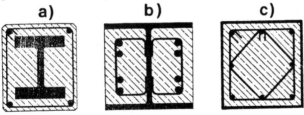

Fig. 12: Typical cross sections of composite columns.

The rules given by ECCS reflect the present state of knowledge. Where appropriate (conservative) assumptions have been made rather than trying to give a detailed description of the phenomena involved. As such, the calculation rules provide for designers an opportunity to benefit from the most recent research results.
Since in some cases only limited experimental evidence is available and/or the theoretical backgrounds are not yet fully understood, the calculation rules are presented in the form of a Technical Note [4].

Thermal Response

ISO standard fire exposure on all sides of the columns is taken as a starting point. By way of reasonable approximation, over the height of the column a uniform temperature distribution is assumed. Under practical fire conditions, however, a significantly non-uniform temperature distribution in the concrete over the cross section must be expected. This is illustrated in Fig. 13 for a concrete filled steel column.

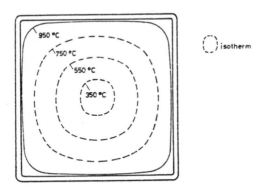

Fig. 13: Measured temperature distribution over the cross section of a concrete filled steel column (200 x 200 x 6.3 m^3) after 90 minutes standard fire exposure on four sides [23].

As a consequence, two dimensional heat flow models must be used. Radiation and convection characteristics of the heat flow are chosen in accordance with the rules given for traditional steelwork.

168

Calculation of the temperature fields over the cross section is only possible by means of a computer. As oposed to the situation with traditional steelwork, no simple geometrical scaling factor like the section factor F/V can be defined. In practice this means that for all relevant cross sectional designs a seperate thermal analysis must be made. In the ECCS-calculation method this work has systematically carried out, as an intermediate step to allow for a proper structural analysis at elevated temperature conditions. A variety of computer programmes is available for this purpose. [24], [25], [26]. A discussion of these is, however, outside the scope of this paper.

Structural Analysis

As already mentioned when discussing the calculation method for traditional steelwork, for columns no simple plastic theory can be used, but an incremental elasto-plastic approach is necessary. Reference is made to Fig. 9.

In the case of composite columns an extra complication arises as a direct consequence of the non-uniform temperature distribution over the cross section. This will cause additional stresses in the cross section due to restrained thermal elongation as exemplified in Fig. 14a. The thermal stresses may have a significant influence on the load bearing capacity of composite columns, as is illustrated by the two buckling curves, presented in Fig. 14b. Both curves are calculated for a reinforced, concrete filled steel column (⊞ 300 x 300 x 7 mm³) after 90 minutes standard fire exposure [27]. The drawn curve is based on an model described in [28]; the dashed curve is obtained for the same situation, however by neglecting the effect of the thermal stresses. Note that, for the given situation, the effect of the thermal stresses for a buckling length of, say, 3 m is almost neglectible. For higher buckling lengths this effect becomes significant.

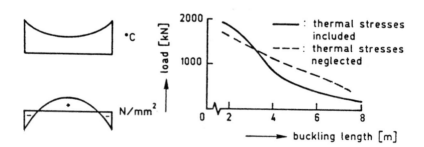

(a) distribution of temperature and thermal stresses (schematic)

(b) buckling curves at elevated temperatures

Fig. 14: The effect of restraint of thermal elongation on the load bearing capacity of fire exposed concrete filled steel columns.

The structural analysis should therefore, ideally, be based on so called "exact" models, i.e. models allowing for a precise thermal and mechanical analysis, cf. e.g. [28], [29].

The numerical complexity of the exact physical models is, however, quickly increasing with growing precision of the analysis. This may have certain drawbacks, especially if design information requires a great number of systematic calculations. For this reason limited, more approximate models have been developed [30], [31]. They are based on assumptions which are similar to those used for the approximate column model for traditional steelwork as described earlier. Ample attention is to be paid to a proper validation of the limited calculation models, which, occasionally, leads to the introduction of semi experimental correction factors. Therefore these methods should be used with caution when extrapolating outside the range of experimental evidence.

The ECCS-Technical Note provides design information in the form of buckling curves for various cross sectional dimensions, profiles and reinforcement, if any, and for periods of standard fire exposure of 30, 60, 90 and 120 minutes. An example of such a design chart is given in Fig. 15.

For columns with steel profiles encased in concrete, the necessary calculations are based on an "exact" model. For the other two invisaged column types (i.e. steel profiles concreted between the flanges and concrete filled steel sections), the ECCS-Technical Note uses "limited" calculation models.

170

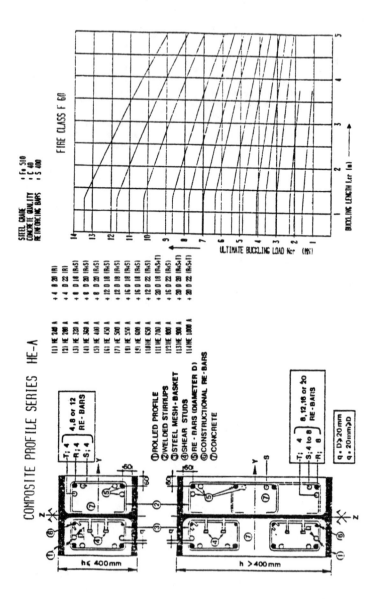

Fig. 15: Buckling curves for various I-profiles with concrete between the flanges after 60 minutes standard fire exposure [4].

171

COMPOSITE SLABS

Fig. 16 shows two typical shapes of composite slabs, dealt with in the ECCS calculation method for fire resistance for composite steel conrete slabs.

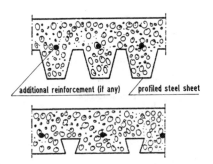

Fig. 16: Typical cross sections of composite slabs.

As opposed to the building components discussed in the preceeding chapters, composite slabs have not only a load bearing but also a seperating function. This means that also the insulation and integrity criterion should be considered when determining the fire resistance. For composite steel concrete floors the integrity criterion is not difficult to fulfill. The main reason is that, normally, the floor slab is cast in situ. This means that joints are adequately sealed. Any cracks which may occur in the concrete during fire exposure are unimportant because the steel sheet will prevent penetration by the flames and hot gases. Therefore it will be assumed that, if the insulation criterion is fulfilled (i.e. temperature rise at the unexposed side should not be over 140 °C on the average, or over 180°C in any point), then the integrity criterion is also fulfilled.

Thus, the following discussion focusses on the ECCS-rules given to analyse the criteria of insulation and load bearing capacity. These rules are presented the form of a Technical Note [5], with a status similar to that of the Technical Note on fire exposed composite columns [4]. As far as the load bearing capacity is concerned, the rules concentrate on the effect of additional reinforcement, since, without such reinforcement the fire resistance of composite slabs is only limited to, say, 30 minutes. Standard fire exposure on the underside only is assumed.

Thermal Response

As for composite columns, also in the case of concrete slabs with profiled steel sheet two dimensional heat flow models are necessary in order to determine the thermal response. For every day design such models must however be regarded as too cumbersome. Therefore, in the Technical Note, the profiled slab is schematized to a flat slab with an

effective thickness equal to a weighted average of the real slab thickness. For various periods of standard fire exposure, temperature distributions are given based on [32]. From such distributions a rule for
minimum slab thickness, necessary to fulfill the criterion of insulation,
can be derived. This rule is given in Fig. 17, together with some test
results. It is seen that, applying the rule, conservative solutions are
obtained.

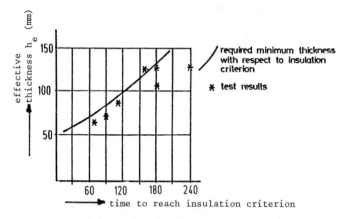

Fig. 17: Minimum effective slab thickness in order to meet the
insulation criterion.

In the structural analysis, the temperature of the addition reinforcement plays a crucial role. The Technical Note gives experimentally
determined equations from which this temperature may be calculated as
function of the position of the reinforcement bar in the slab, given by
u_1, u_2, u_3 and the period of standard fire exposure, t (cf. Fig. 18a).
Fig. 18 b shows the results of some validation tests.

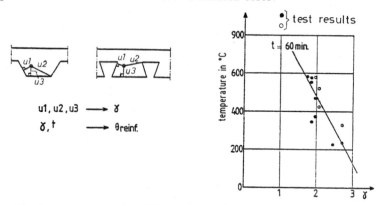

Fig. 18: Temperature in the additional reinforcement.

Structural Analysis

The load bearing capacity is determined on basis of simple plastic theory. This requires in general the quantification of both a positive and a negative plastic bending moment at a given period of standard fire exposure. Cf. Fig. 19.

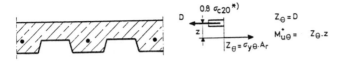

$$M^+_{u\theta} + M^-_{u\theta} \leq q \cdot L^2/8 \qquad \text{or}$$

$$q \geq 8 \cdot (M^+_{u\theta} + M^-_{u\theta})/L^2$$

Fig. 19: Determination of the load bearing capacity of composite slabs according to the elementary plastic theory.

The positive and negative bending moments denoted by $M^+_{p\theta}$ and $M^-_{p\theta}$ respectively, can be evaluated on basis of the typical stress distributions as presented in Fig. 20 and 21.

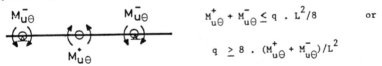

Fig. 20: Positive plastic moment $M^+_{u\theta}$

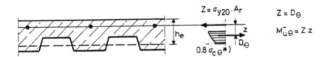

Fig. 21: Negative plastic moment $M^-_{u\theta}$

The following simplifying assumptions are made:

General:
- the tensile strength of concrete does not contribute to the load bearing capacity at elevated temperatures and thus may be ignored;
- the steel sheet does not contribute to the load bearing capacity at elevated temperatures and thus may be ignored.

For the positive plastic moment:
- the ultimate strength of concrete in the compression zone is not influenced by temperature and room temperature values may be taken;

----

*) The factor 0.8 is introduced to correct for the assumed full plastic stress distribution in the concrete compression zone. In the ultimate state, a non uniform stress distribution will occur, due to the limited capacity of concrete to accept deformation.

- the effective yield stress of the additional reinforcement is affected by temperature; this temperature is a function of the position of the reinforcement bars and period of fire exposure, as discussed earlier.

For the negative plastic moment:
- in calculations, the profiled concrete slab may be replaced by a slab with a uniform thickness equal to the effective thickness as used in the thermal analysis;
- the ultimate strength of concrete in the compression zone (exposed side) is affected by the temperature;
- the effective yield stress of the reinforcement (unexposed side) is not influenced by temperature and room temperature values may be used.

On basis of the above mentioned assumptions, the evaluation of the failure conditions can proceed in a similar way as for conventional reinforced concrete slabs under ambient temperature conditions. It is also possible to construct design tables or diagrams from which the plastic moments can be directly read as function of the relevant parameters.

ACKNOWLEDGEMENT

The calculation methods discussed in this paper have been elaborated by Technical Committee 3 "Fire Safety of Steel Structures" of the European Covention for Constructional Steelwork. The stimulating discussion within this Committee is acknowledged here.

REFERENCES

[1] J. WITTEVEEN: "A systematic Approach Towards Improved Methods of Structural Fire Engineering Design". Proceedings 6th International Fire Protection Seminar organized by VFDB, Karlsruhe, 1982.

[2] CIB/W14: "A Conceptional Approach Towards a Probability Based Design Guide on Structural Fire Safety". Fire Safety Journal, Vol. 6, No. 1, 1983.

[3] EUROPEAN CONVENTION FOR CONSTRUCTIONAL STEELWORK: "European Recommendations for the Fire Safety of Steel Structures: Calculation of the Fire Resistance of Load Bearing Elements and Structural Assemblies Exposed to Standard Fire". Elsevier, Amsterdam—Oxford—New York, 1983.

[4] EUROPEAN CONVENTION FOR CONSTRUCTIONAL STEELWORK: "Calculation of the Fire Resistance of Composite Columns Exposed to the Standard Fire". To be published 1st half year 1986.

[5] EUROPEAN CONVENTION FOR CONSTRUCTIONAL STEELWORK: "Calculation of the Fire Resistance of Composite Concrete Slabs with Profiled Steel Sheet Exposed to the Standard Fire". Publication No. 32, Brussels, April 1984.

[6] ISO: "Fire Resistance Tests - Elements of Building Construction". International Standard 834, 1975.

175

[7] EUROPEAN CONVENTION FOR CONSTRUCTIONAL STEELWORK: "Design Manual on the European Recommendations for the Fire Safety of Steel Structures. Publication No. 35, Brussels, 1985.

[8] W. GEILINGER and S. BRYL: "Feuersicherheit der Stahlkonstruktionen" IV-Teil, Schweizer Stahlbauverband, Zürich, Heft 22, 1962.

[9] J. WITTEVEEN: "Brandveiligheid staalconstructies (Fire Safety of Steel Structures)", Stichting Centrum Bouwen in Staal, Rotterdam 1965. Revised and extended edition by L. Twilt and J. Witteveen, Staal Centrum Nederland, 1980.

[10] S.E. MAGNUSSON, O. PETTERSSON and J. THOR: "Fire Engineering Design of Steel Structures". Publication 50, Swedish Institute of Steel Construction, Stockholm 1976 (Swedish version 1974).

[11] CENTRE TECHNIQUE INDUSTRIEL DE LA CONSTRUCTION METALLIQUE: "Methode de Prévision par le Calcul du Comportement au feu des Structures en Acier", Construction Métallique No. 4, 1976.

[12] FIRE SAFETY IN CONSTRUCTIONAL STEELWORK: CECM-III-74-2E. European Convention for Constructional Steelwork 1974.

[13] N.W. STANZAK, T.T. LIE: "Fire Resistance of Unprotected Steel Columns". Journal of the Structural Division, ASCE, ST5, May 1973.

[14] TEESIDE LABORATORIES: "The Fire Resistance of Unprotected Steel Beams". Doc. ISO/TC92/WG15 N13 British Steel Corp., September 1980.

[15] J.H. Mc GUIRE, N.W. STANZAK and M. LAW: "The Scaling of Fire Resistance Problems". Fire Technology, Vol. 11, No. 3, August 1975.

[16] M. LAW: "Monograms for the Fire Protection of Structural Stelwork". Fire Prevention Science and Technology, No. 3, November 1972.

[17] J. WITTEVEEN and L. TWILT: "Behaviour of Steel Columns under Fire Action". International Colloquium on Column Strength, Paris 1972, Proceedings IABSE, Vol. 23, Zürich 1975.

[18] J. WITTEVEEN, L. TWILT and F.S.K. BIJLAARD: "The Stability of Braced and Unbraced Frames at Elevated Temperatures". Second International Colloquium on the Stability of Steel Structures, Liège 1977, Preliminary Report.

[19] J. KRUPPA: "Résistance au Feu des Structures Métallique en Temperature Non Homogène". Thesis l'Institut National des Sciences Appliquées de Rennes, France 1977.

[20] J. JANSS and R. MINNE: "Buckling of Steel Columns in Fire Conditions". Fire Safety Journal, Vol. 4, No. 4, 1981/1982, p.p. 227-235.

[21] J. RONDEL and R. MAQUOI: "Formulations d'Ayrton-Perry pour le Flambement des Barres Métalliques". Construction Métallique No. 4, 1979.

176

[22] O. PETTERSSON and J. WITTEVEEN: "On the Fire Resistance of Structural Steel Elements Derived from Standard Fire Tests or by Calculation". Fire Safety Journal, Vol. 2, 1979/1980, pp. 73-87.

[23] K. KORDINA and W. KLINGSCH: "Brandverhalten von Stahlstützen im Verbund mit Beton und von massiven Stahlstützen ohne Beton". Forschungsbericht P35, Studiengesellschaft für Anwendungstechnik von Eisen und Stahl e.v. Düsseldorf/EGKS 7219 SAI/108, 1984.

[24] U. WICKSTROEM: "TASEF/2, A Computer Programme for Temperature Analysis of Structures Exposed to Fire". Lund Institute of Technology, Rep. No. 79,2 1979.

[25] R. RUDOLPH and R. MULLER: "ALGOL/Computerprogramm zur Berechnung zweidimensionaler instationärer Temperaturverteilungen mit Anwendungen aus dem Brand- und Wärmschutz". BAM, Forschungsbericht 74, Berlin, 1980.

[26] SAMCEF: "Systéme d'Analyse de Milieux Continus par Elements Finis". Laboratoire de Techniques Aéronautiques et Spatiales, LTAS, Université de Liège, 1982.

[27] L. TWILT and P.W. v.d. HAAR: "The Effect of the Mechanical Properties of Concrete and the Thermal Induced Stresses on the Discrepancy between the French and German Calculation Models for the Fire Resistance of Concrete Filled Steel Columns". TNO-report No. B-85-93, March 1985.

[28] U. QUAST, R. HASS and K. RUDOLPH: "STABA/F, A Computer Programme for the Determination of Load Bearing and Deformation Behaviour of Uni-Axial Structural Elements under Fire Action". Technical University of Braunschweig, March 1984.

[29] J.B. SCHLEICH, J.C. DOTREPPE and J.M. FRANSSEN: "Numerical Simulations of Fire Resistance Tests on Steel and Composite Structural Elements or Frames". Report presented at the FIRST INTERNATIONAL SYMPOSIUM ON FIRE SAFETY SCIENCE, Gaithersburg, Maryland, USA, October 1985.

[30] W. KLINGSCH: "KSTTR, Computer Programme for Load Bearing Analysis of Steel, Reinforced Concrete and Composite Columns in Fire Case (Physical and Geometricla Non-Linear)". SFB 148, Technical University of Braunschweig, 1975.

[31] G. GRANDJEAN, J.P. GRIMAULT and L. PETIT: "Determination de la Durée au Feu des Profils Creux Remplis de Béton". Cometube, 1980.

[32] "FIB/CEB Report on Methods of Assesment of the Fire Resistance of Concrete Structural Members". FIP Commission on the Fire Resistance of Prestressed Concrete Structures, 1978.

THE DESIGN OF STRUCTURES AGAINST FIRE
SOME PRACTICAL CONSTRUCTIONAL ISSUES

SUTHERLAND, R.J.M

Partner, Harris & Sutherland
Consulting Civil & Structural Engineers
London

ABSTRACT

Fire precautions should be considered creatively as a design matter. Today there is confusion between safety and the preservation of property. Do structural fire resistance periods relate at all to safety and could they be reduced possibly given other precautions? Is thermal movement adequately considered? How could exposed steel, timber and cast iron be made more fire-resistant without losing visual appeal. These are the sort of practical questions which trouble designers searching for quality and thwarted by a doctrinaire attitude to fire.

INTRODUCTION

I am writing this paper very much from the angle of a general structural designer who at times has been frustrated by fire regulations and at others puzzled by what apparently works or is acceptable. If some of the questions I ask, or the suggestions I make, seem naive to the cognoscenti of the fire world I apologise but console myself with the thought that many share my uncertainties.

Architects and engineers tend to look on fire precautions in the same way as they look on income tax, that is as a necessary evil to be satisfied or got around with the minimum impact on other requirements. They try to "persuade the fire officer ....." This common opening to any discussion on the subject sums up the attitude. Few try to design creatively for fire.

Some of the negative attitude to fire derives from ignorance, some from emotion - horror at disasters like Bradford - and some from a feeling that it is someone else's problem. "We will be told what to do."

Keeping means of escape on one side there is confusion between a structural fire resistance period, resistance to flame spread and incombustibility. Our forefathers saw incombustibility as the answer to all problems. They converted their mills from timber to iron with brick arches, by and large with good results, yet using present day criteria the timber may well have had an appreciably longer fire resistance period than the iron. Perhaps our forefathers were not so far wrong.

There is also confusion in codes and regulations between health and safety on one hand and the protection of property on the other. The Building Regulations are intended to cover no more than health and safety (apart from the relatively recent and illogical inclusion of energy conservation), yet much of the thinking behind the fire clauses seems to have derived only from a desire to preserve property.

It may be because of these various levels of confusion that designers become frustrated and stop thinking about fire, or never start. The aim of this paper is to look at some of these points of uncertainty or frustration and see whether a more positive attitude to fire precautions would be possible. Of course there is the advanced concept of full fire engineering, where fire precautions, fire load and actual use are closely allied, but I am not proposing to go quite as far as that. I intend to concentrate on the more niggling little doubts and aggravations which affect everyday design today, starting with some generalities and then looking at the problems of individual materials.

FIRE RESISTANCE PERIODS

How were these derived and why are they so long? Clearly they have little to do with the safety of the occupants of a building and it is hard to visualise the circumstances in which the difference between one hour and two would be vital for the safety of firemen, let alone the difference between two hours and four.

If the longer fire periods - or any specified periods - are not needed for the safety of people should they be deleted from the Building Regulations and treated as an insurance matter? If preservation of property is the criterion, would sprinklers often be more effective and also cheaper when insurance premiums are taken into the calculation?

The cost of sprinklers would need to be considered as well as their effectiveness. Sprinklers cost money and need maintenance while an in-situ concrete floor will generally cost more but have a better fire resistance than a timber one; the problem is to find a balance.

The question whether insurance alone could be used to ensure adequate safeguards for people is worth debating. Possibly some minimum structural fire resistance periods would still be desirable within buildings but at a more liberal level than now, say $1/2$ to 1 hour, or even less, instead of 1 to 2 hours.

With fire, as with sound insulation, it seems that in practice it is often the unsuspected bypassing of the resistant construction which causes trouble rather than the level of resistance of the materials used. With most forms of construction, especially in the case of existing buildings, a reduction in the fire resistance periods, if justifiable, would give very real benefits as I will try to show.

With vertical separations of ownership or occupancy there may be a case for tighter control but even here it is worth questioning whether any but the riskiest occupancies would merit a separation of more than one hour. It might be better to have more vertical barriers each with a lower time rating.

ROOFS

It has been suggested that there should be a structural fire resistance period, say a quarter of an hour, for all roofs. The first question is whether this would in any way increase safety. Such a requirement could quickly lead to absurdities. For instance is a glazing bar a structral element?

Some authorities have asked for a fire resistance period for roofs well in excess of a quarter of an hour. For instance, in the case of the Commonwealth Institute in Kensington, 25 years ago the District Surveyor insisted on two hours for the main roof. This is a public building and he was quite within his rights but his action reduced the number of design options sharply. One can sympathise with his concern for safety in a large open space for public assembly but I am not convinced that he improved this to any real, extent with a two-hour structure. If calculated in the same way the Dome of Discovery built only a few years earlier must have had almost no fire resistance.

The case for a fire resistance period for roof structures on grounds of safety does not seem to be real except possibly in situations where failure of part of a roof as a result of a fire could lead to a collapse well away from the area initially affected. For instance if the failure of one or two roof trusses in a long storage building led to a ripping action which pushed over the walls this could be positively unsafe and give virtually no warning even immediately before collapse. However this would be a broader case of instability, not confined to fire, and thus best looked on as a design matter and certainly not one for a prescriptive fire resistance period.

Even if there is no case for a structural fire period for roofs, on grounds of safety, there may be an economic one related in part to insurance. I will return to this is connection with a number of roof details, especially with timber, but first a quick look at concrete and steel.

CONCRETE

Here for some time the big problem with fire has undoubtedly been supplementary mesh or other reinforcement as required by CP 110. The near impossibility of maintaining this mesh in place has become less of a problem with the issue of BS 8100 which effectively confines the need for it or equivalent precautions to fire resistance periods of more than two hours. However this prompts me again to ask when and where are fire resistance periods greater than, or as great as, two hours really needed.

Given an overall reduction in fire resistance periods any doubts on the adequacy of trough and waffle floors should evaporate.

It would be instructive to analyse actual major fires in buildings with in-situ reinforced concrete frames, whether positively designed for continuity or not, and calculate the percentage of collapses. I suspect that the figure would be very small.

There may be a case for distinguishing between in-situ and precast concrete when considering fire, not because the place of casting affects the fire resistance of the actual material but because the tying together of the elements may be less robust with precast work. there could also be more problems with precast than with in-situ concrete due to thermal expansion and contraction in fires.

It is seldom that thermal movements can be measured in a real fire. This is where the evidence of the well-monitored BRE/BDP test in a flat in Ronan Point is so valuable. In a fire which peaked $1000^{\circ}C$ in a period of 15 minutes a precast slab 5 metres long expanded 7mm. with a recorded temperature rise within the concrete of $150^{\circ}C$. This implies a coefficient of thermal expansion of 9.3×10^{-6} per $^{\circ}C$. The accepted figure for steel is 12×10^{-6} and for concrete $9-14 \times 10^{-6}$ which indicates that not much of the thermal expansion was counteracted by any restraint. Whether such a movement is likely to cause more than a problem of serviceability is debatable but given five times the heated length - by no means impossible with either in-situ or precast concrete - the movement would be 35mm which could cause more serious disruption unless allowed for; what is more the disruption could well occur at a point well away from the seat of the fire.

The question which this raises is whether in some circumstances we should incorporate fire expansion joints to limit the damage to property. I would do so with reluctance. Movement joints provided to allow for normal thermal and moisture movements often cause more problems than they solve. However the point is worth considering.

STRUCTURAL STEEL (THERMAL MOVEMENT)

If thermal expansion can cause problems with reinforced concrete in a fire how much more so with structural steel especially in exposed roof structures.

Figure 1 shows the plan of a three storey "heat recovery" teaching block in a further educations college, all of reinforced concrete up to 2nd floor level and with a steel roof measuring 50.4m x 42m supported on concrete columns and stabilised by four service towers. A fire started in the upper floor at the point marked X and in time the whole roof collapsed as did a large part of the vertically reinforced brick cladding to the top floor, the rest being badly damaged.

Assuming that steelwork loses its strength at 550°C it is very possible that an average temperature rise of 300°C occurred before anypart of the structure collapsed. If so the expansion would have been about 180mm, quite enough to disrupt the cladding. One cannot tell how much of this disruption was due to thermal expansion and how much to the final collapse of the steelwork but, having seen the problem, it would not be difficult to make some "weak links" to allow either movement in the roof structure and save the walls.

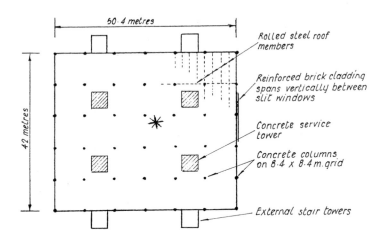

Figure 1 : PLAN OF FURTHER EDUCATION COLLEGE
BLOCK SHOWING TOP FLOOR & ROOF
CONSTRUCTION

Perhaps the most significant aspect of this fire is that the damage would probably have been confined to the top floor if no attempt had been made to put out the fire. As it was the carpet on the top floor largely survived the fire but all the services, fittings and finishes throughout the building were ruined by water.

Commercially this experience points to the desirability of vertical fire breaks and a fire fighting policy based on controlling spread rather than necessarily extinguishing fires.

STRUCTURAL STEEL VISUALLY EXPOSED

Many architects would like to use more exposed steelwork both within buildings and outside. Given neat detailing, steelwork has a crispness and solidity which is hard to match, a quality currently much admired in 19th Century structures and one accentuated by the current flight from visible concrete.

182

Most casings seem to be either flimsy or bulky, shielding restricts
the layout and may lead to problems of weathering, while intumescent
coatings tend to look rough and are easily damaged. Concrete filling
of tubes makes solid looking columns a possibility but there is the
problem of how to join onto them.

Figure 2 shows two possible junctions. With 2(a) the fire-proofing is
certain but the appearance is a matter of taste. With 2(b) the
vertical line of the column is preserved and watertightness can be
ensured by welding but the fire resistance is perhaps less certain.

Take the detail one step further and eliminate the inside joist in
Figure 2 (b) and then what happens to the fire resistance?

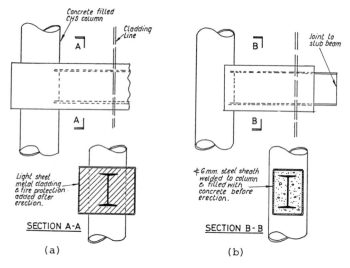

Figure 2: POSSIBLE BEAM/COLUMN JUNCTIONS
FOR EXTERNAL STEEL FRAME

This brings one to the question of horizontal spans with filled steel
tubes or channels. Possible sections are shown in Figure 3.
Construction could have the speed of steelwork, and its elegance, but
what fire resistance period could be achieved - or is really needed?
With all of these reinforcement for top of the slab continuity at
the supports would clearly be benefical.

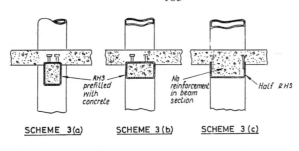

SCHEME 3(a)   SCHEME 3(b)   SCHEME 3(c)

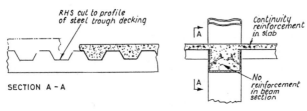

SECTION A - A

SCHEME 3(d)

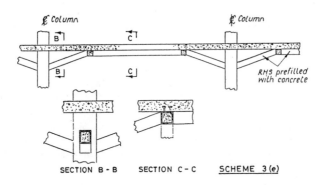

SECTION B - B   SECTION C - C   SCHEME 3(e)

Figure 3 : POSSIBLE (?) BEAM FORMS WITH
EXPOSED STRUCTURAL STEEL

Am I asking too much?   There are some encouraging signs.

Fire resistance periods of up to at least an hour have been
demonstrated in tests on steel trough floors with concrete topping but
with no reinforcement. Presumably their survival depends on the rate
at which heat is conducted away from the steel but it is difficult to
get solid information.

Perhaps the most surprising "success" story, again it seems depending on conduction of heat away, is that of interlocking channel reinforced woodwool. If this can have a two-hour fire certificate, based on 75 mm thickness spanning 3 metres under its normal design load with the steel exposed below, surely there is scope for more adventurous uses of steel in other fields.

SEPARATING SAFETY & SERVICEABILITY

An aspect of design frequently ignored by engineers, in spite of limit state codes, is the desirability of thinking of ultimate strength separately from everyday performance. For instance a floor needs to feel rigid when walked upon but in the ultimate limit state all it needs to do is to stand up.

Good use of this distinction was made recently with the galleries of a community theatre. Here it proved possible to extend quite thin reinforced concrete floors as 'ultimate cantilevers' yet use light steelwork screening with no provable fire resistance to prop them so as to feel comfortable to users.

This principle could well be extended beyond the case of the simple cantilever. Beams of slabs could even be designed to survive as catenaries in fire yet be stiffened by slender exposed steel columns or elegantly designed steel trussing for normal conditions. The step from a continuous beam to a catenary is only a small one. This thought leads on to the cast iron column, and cast iron in general, in fire.

CAST IRON

There is a constant succession of buildings coming up for conversion where the fire resistance of cast iron acts as a barrier to what designers wish to do.

Recent tests sponsored by the GLC and described by Barnfield & Porter in a very interesting paper published by the ISE (1) go some way towards clearing some of the doubts about cast iron columns in fire. Brittleness is - predictably - the main problem together the effects of distortion. This is obviously a case where disruption by thermal movements in fire should be limited and what better way to achieve this, where practicable, than by using timber beams on cast iron columns rather than steel or iron ones. However, with intumescent coatings on the exposed parts of cast iron beams and the columns fully exposed, the performance of an all iron system could possibly be improved greatly. More testing is certainly need here.

Most engineers look on cast iron columns only as a possible means of providing ultimate support. They could also be used for stiffness in situations, as described in the last section, where survival depends on some other structural action.

185

Whether used for ultimate support or just to provide stiffness, one cannot help wondering whether the brittleness of cast iron columns and the effect of diplaced cores could not be reduced by injection with grout or, where possible, by filling with fibre-reinforced concrete. Given access perhaps a vertical steel bar encased in grout or concrete (as Figure 4) would prevent a brittle collapse - another possible line for research.

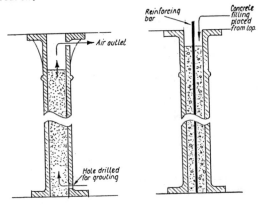

Figure 4 : POSSIBLE MEANS OF REDUCING BRITTLENESS OF CAST IRON COLUMNS

So far I have been looking at cast iron columns and beams as historical survivals which we wish to continue to keep. I do not dismiss the possibility that given adequate precautions against brittle failure in fire, cast iron columns might be used from scratch in new buildings. They could again be provided with decorative details, equivalent to those used in the 19th Century, but still look and feel solid and durable.

TIMBER IN ROOFS

Perhaps it is because there is no requirement for fire resistance to be considered in roofs that so little attention has been given, or so little published, on the performance of timber connectors in fire.

Reputedly nail plates have been shown to fail at as little as $7^1/2$ minutes in a fire (confirmation of this seems impossible). The timber sections which they join tend to be about 97 x 41. Allowing for charring at 40 mm per hour [2] one would expect the timber to continue to carry its load for at least twice times as long. Trussed rafters are normally protected by ceilings, but not always. Might it not make commercial sense to market simple protective covers which could just be hammered on (Figure 5a). Apart from anything else the rafters would look much better, given that the covers were large enough to be square to the edges - unlike many nail plates!

186

The same thinking could be applied to bolted connections, perhaps with even greater advantage. Again, given some thought to their design, the protective covers could greatly improve the look of the trusses; make them look more neat and robust (Figure 5b).

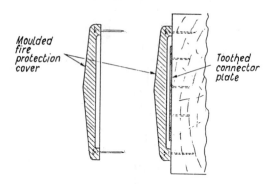

Figure 5(a) : POSSIBLE FIRE PROTECTING COVER
FOR TOOTHED CONNECTOR PLATES

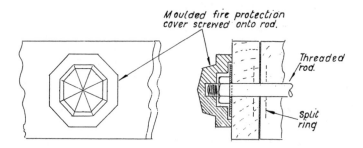

Figure 5(b) : POSSIBLE FIRE PROTECTING COVER
FOR BOLTED TIMBER TRUSS CONNECTION

The point with such precautions is that, if standard products were available, the extra initial costs should be small whereas in a limited fire the difference in cost between local repair and replacement after collapse could be large.

CONCLUSION.

I have tried to indicate by example some ways in which the ordinary engineer's horizons might be broadened and his attitude to fire become more positive and creative. To make this possible there is a need to remove some of the mystery from the subject, to make information from tests more widely available, to carry out more tests and above all to discuss the problems in real terms rather than in relation to existing regulations. Perhaps this conference will go further than anything organised previously towards achieving a more enlightened attitude. I certainly hope so and applaud the foresight of the organisers.

REFERENCES

(1)   Barnfield, J.R. & Porter, AM. : 'Historic buildings and Fire: fire performance of cast-iron structural elements' The Structural Engineer 62A No.12 December 1984.

(2)   B.S 5268 : Part 4.1 : 1978

DESIGN IMPLEMENTATION — CONCRETE STRUCTURES

ARMAND H. GUSTAFERRO

The Consulting Engineers Group Inc.
1701 E. Lake Avenue
Glenview, Illinois 60025
U. S. A.

HISTORY

In the past, architects, engineers, building officials, and the construction industry have relied almost entirely on results of standard fire tests to evaluate the fire resistance of building components. Until about 25 years ago, little was done to "design" structures to resist fire as we design structures to resist the effects of gravity, wind, and earthquakes. More recently, a new field is emerging, which Professor Ove Petterson of Sweden calls, "fire engineering design of buildings" (1)*.

The standard fire test method used in most countries is ISO 834. In the United States, we use ASTM E119 which is essentially the same as ISO 834 insofar as it is applied to fire tests of concrete assemblies. Basically the standards require that a specimen of a wall, column, beam, floor, or roof be exposed to a "standard" fire. The standard fire is defined in terms of a time-temperature relationship. During fire tests, load-bearing elements must support loads which simulate the effects of service loads. The fire endurance of a specimen is the duration of a fire test until an end point is reached.

The basic end point criteria are:

1) the specimen must support its service loads—collapse is an obvious end point.
2) passage of flame or hot gases must not occur.
3) the temperature of the unexposed surface must not increase more than an average of 140°C or a maximum at any point of 180°C.

These end points are often referred to as stability, integrity, and insulation, respectively.

The discussion which follows deals only with the first criterion.

Results of fire tests give little insight into the structural behavior of specimens during the tests. To develop such insight, we can draw on our experience with other types of tests. For example, as structural engineers we

---

* Numbers in parenthesis indicate references listed at the end of the paper.

understand the behavior of a simply supported beam during a load test. The beam has a certain moment capacity and shear capacity. Those capacities remain constant during a load test. As the loads are increased the applied moments and shears are increased. If the beam fails in bending rather than in shear, we know that the applied moment has reached the moment capacity.

During a fire test of the same simply supported beam, the applied load remains constant. As the test progresses, the moment capacity and shear capacity diminish. If the beam fails in bending, we can assume that the moment capacity has been reduced to the magnitude of the applied moment.

Even though the above statements appear to be obvious, very few tests were conducted before about 1960 to determine if the theory was correct. Fortunately within the past 25 years, a great deal of research has been conducted on the behavior of structures subjected to fire. This research has led to the development and refinement of methods for calculating the fire resistance of structural elements. This does not mean that all of the needed research has been completed. To the contrary, we have made a beginning but there are still many unanswered questions.

PROPERTIES OF STEEL AND CONCRETE AT HIGH TEMPERATURES

Fire engineering design methods require data on physical properties of steel and concrete at high temperatures.

Fig. 1 shows strengths of three types of steel at elevated temperatures. Data for hot-rolled steel represents a lower bound of the yield strength for hot-rolled structural steels and hot-rolled reinforcing bars (2). Data for cold-drawn wire or strand were developed by Abrams and Cruz (3). High strength alloy bars are the type used for prestressing and the data apply to ultimate strength (4).

Fig. 2 shows data from the U. S. Steel Design Manual (2) on modulus of elasticity of steel at high temperatures.

Fire tests of concrete elements indicate that concretes can be classified by aggregate type and unit weight. In America, we classify four general types of structural concrete: 1) siliceous aggregate concrete, 2) carbonate aggregate concrete, 3) sand-lightweight concrete, and 4) lightweight aggregate concrete.

Siliceous aggregate concretes are made with normal weight aggregates consisting mainly of silica or compounds other than calcium or magnesium carbonate.

Carbonate aggregate concretes are made with aggregates consisting mainly of calcium or magnesium carbonate, e.g., limestone or dolomite.

Sand-lightweight concretes are made with a combination of expanded clay, shale, slag, or slate or sintered fly ash and natural sand. Their unit weights are generally between 1700 and 1950 kg/m (3).

Lightweight aggregate concretes are made with aggregates of expanded clay, shale, slag, or slate or sintered fly ash, and weigh about 1350 to 1850 kg/m (3).

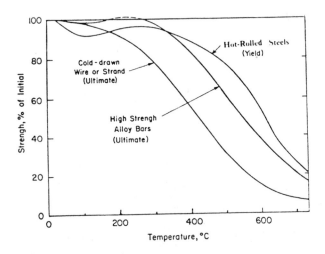

Fig. 1 — Strength of certain types of steel at high temperatures

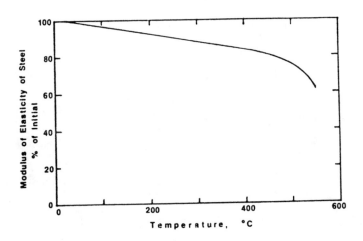

Fig. 2 — Modulus of elasticity of hot-rolled steel at high temperatures

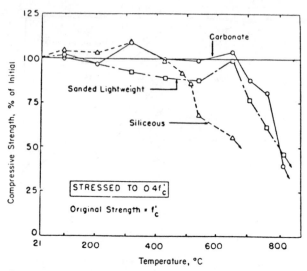

Fig. 3 — Compressive strength of structural concrete at high temperatures

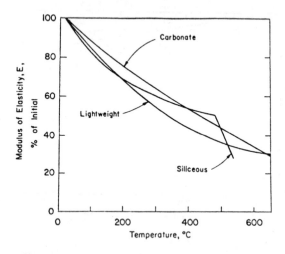

Fig. 4 — Modulus of elasticity of concrete at high temperatures

Siliceous and carbonate aggregate concrete are also considered to be normal weight concretes.

Fig. 3 shows data developed by Abrams (5) on the compressive strength of concrete at high temperatures. The data apply to concrete that is in compression during the heating period. During the heating period cylinderical specimens were stressed 0.4 of their room temperature strength (0.4 f_c'). Other tests conducted with specimens stressed to 0.25 f_c' and 0.55 f_c' gave approximately the same results.

The modulus of elasticity of concrete is reduced significantly at high temperatures, as shown in Fig. 4 (6).

Creep of concrete is significantly affected at high temperatures as shown in Fig. 5.

Fig. 6 shows thermal expansion of steel and various types of concrete at high temperatures (8, 9).

TEMPERATURES WITHIN CONCRETE ELEMENTS DURING FIRE TESTS

Because properties of materials are temperature-dependent, it is necessary to know the temperatures within structural elements during fire exposure. For calculations involving fire resistance ratings, the fire exposure is that required for standard fire tests.

Figs. 7, 8, and 9 show the temperatures within concrete slabs made with siliceous, carbonate, and sand-lightweight aggregate concretes (10).

Temperatures within concrete beams and joists are more difficult to depict, because they are affected by joist width and aggregate type. Fig. 10 (11) shows the isotherms at 1 hr within a 304-mm wide beam exposed to a standard fire from the sides and bottom. A voluminous compendium of such isotherm drawings would be needed to cover the usual sizes and concrete types for exposure periods of 1, 2, 3, and 4 hours. Furthermore, interpolating among such drawings is laborious. Some computer programs are available for estimating temperatures within beams and joists, but I have found that graphs such as those in Fig. 11 (10) are adequate for most problems.

Fig. 11 shows the temperatures along the vertical centerline of beams or joists at 2 hr of standard fire exposure. From the data in such graphs, one can sketch the isotherms within the beam or joist with adequate accuracy by shaping the isotherms so they look like the ones shown in Fig. 10.

BEHAVIOR OF FLEXURAL MEMBERS DURING FIRE TESTS

Simply supported slabs or beams

Consider a simply supported reinforced concrete slab subjected to fire from below. Assume that the reinforcement consists of straight bars located near the bottom of the slab. With the underside of the slab exposed to fire, the bottom will expand more than the top, and the slab will deflect. Also the strength of the

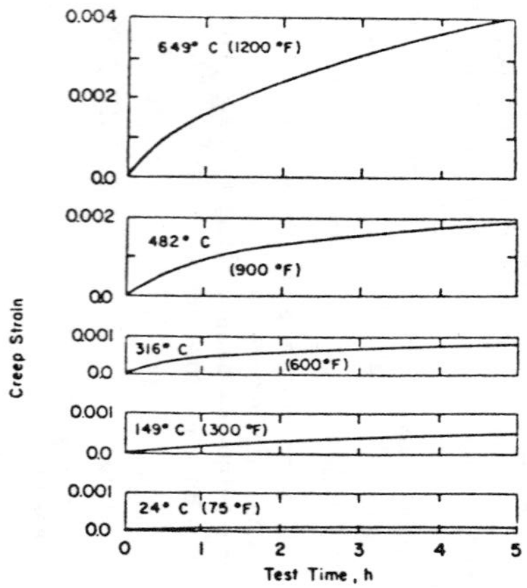

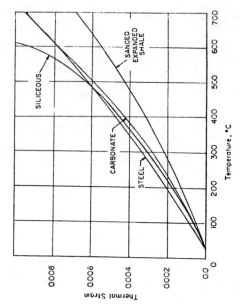

Fig. 6 — Thermal expansion of concrete and steel at high temperatures

Fig. 5 — Creep of carbonate aggregate concrete stressed to 0.45 f'_c

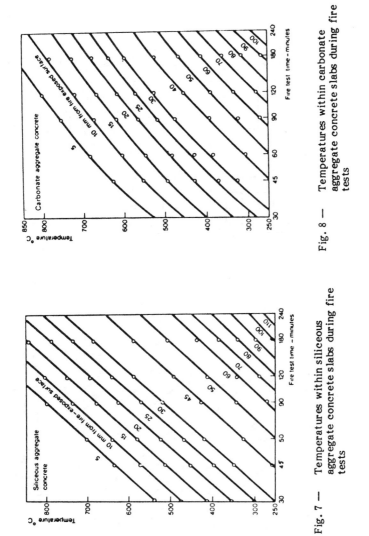

Fig. 8 — Temperatures within carbonate aggregate concrete slabs during fire tests

Fig. 7 — Temperatures within siliceous aggregate concrete slabs during fire tests

196

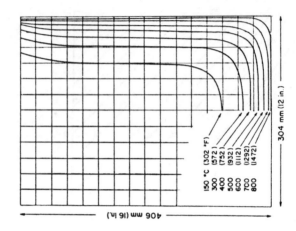

Fig. 10 — Temperatures within a 304-mm wide beam at 1 hr of fire exposure. The beam was made of normal weight concrete

Fig. 9 — Temperatures within sand-lightweight concrete slabs during fire tests

197

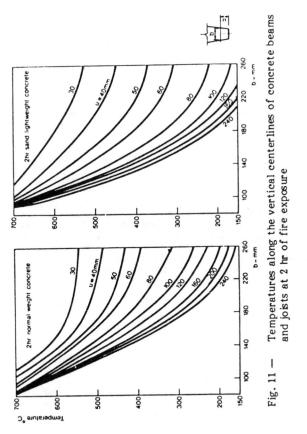

Fig. 11 — Temperatures along the vertical centerlines of concrete beams
and joists at 2 hr of fire exposure

concrete and steel near the bottom of the slab will decrease as the temperature increases. When the moment capacity of the section is reduced to that of the applied moment, flexural collapse will occur.

It is apparent from the above description that the steel temperature at which collapse occurs depends on a) the ratio of applied moment to moment capacity and b) the type of steel.

A series of fire tests of simply supported prestressed concrete slabs (12) verified the above theory. These tests also demonstrated that fire endurance (as determined by stability) of simple slabs could be predicted with reasonable accuracy by applying structural engineering principles together with data on properties of materials at high temperatures and temperatures within slabs caused by fire exposure. Subsequent tests on simply supported beams (4) further confirmed the theory.

Example A, attached, illustrates the calculation methods for a simply supported reinforced concrete slab. Note that the formula for theoretical moment capacity during fire exposure $M_{n\theta}$, is essentially the same as the formula for normal temperatures. The only differences are that the yield strength of the steel, f_y, and the depth of the equivalent rectangular stress block, a, must be modified for the temperatures within the slab at 2 hr of fire exposure. In this case, the strength of the concrete within the compressive stress block is not reduced because the temperature of the concrete is well below $250^\circ C$ (Figs. 7 and 3).

For prestressed concrete with bonded reinforcement, the stress in the steel at ultimate should be calculated, but for most cases it is 96 to 99% of the ultimate strength of the steel. If however the steel is unbonded, the stress at ultimate must be calculated because it is significantly lower than the ultimate strength.

Continuous slabs and beams

Structures that are continuous or otherwise statically indeterminate, undergo changes in stresses when subject to fire. It should be noted that this is different than simply supported members where the applied moments at a section remain constant during fire exposure.

Consider a two-span continuous reinforced concrete slab with rocker-rollers at the outer supports. During fire exposure from beneath, the underside of the slab expands more than the top. This differential heating causes the ends of the slab to tend to lift from the outer supports thus increasing the reaction at the interior support. This action results in a redistribution of moments, i.e., the negative moment at the interior support increases while the positive moments decrease.

During the course of a fire, the negative moment reinforcement remains cooler than the positive moment reinforcement because it is better protected from the fire. Thus, the increase in negative moment can be accommodated. The resulting decrease in positive moment means that the positive moment steel can withstand a higher temperature before failure will occur. Thus, the fire endurance of a continuous member is generally significantly longer than that of a simply supported member having the same cover and load intensity.

Fire tests conducted at the Portland Cement Association (3), in Germany (14), and in Holland (15) have verified the theory outlined above.

Shear strength of concrete beams

Little research has been done in the U.S.A. on the shear resistance of concrete beams subjected to fire. Based on observations during fire tests of beams and slabs, shear failures were indeed rare. In fact, I am not aware of any shear failures in simply supported beams or slabs during fire tests. We observed one shear failure in a continuous beam, but that was in a beam which was inadequately reinforced for shear at normal temperatures. It should be noted that in continuous beams or slabs, a redistribution of moments can be accompanied by a redistribution of shear particularly near the first interior support, where shear and moment can increase dramatically.

In Germany, Krampf reported on a series of 27 fire tests directed toward the study of shear resistance of reinforced concrete beams exposed to fire (16). The report concludes that simply supported beams are not very susceptible to shear failures during fire exposure. Two-span continuous beams are more vulnerable to shear failures so adequate shear reinforcement should be provided near interior supports.

Members in which restraint to thermal expansion occurs

If a fire occurs beneath a small interior portion of a large reinforced concrete slab, the heated portion will tend to expand and push against the surrounding part of the slab. In turn, the unheated part of the slab exerts compressive forces on the heated portion. The compressive force, or thrust, acts near the bottom of the slab when the fire first occurs, but as the fire progessses the line of action of the thrust rises as the heated concrete deteriorates. If the surrounding slab is thick and heavily reinforced, the thrust forces that occur can be quite large, but are considerably less than that calculated by use of elastic properties of concrete and steel together with appropriate coefficients of expansion. At high temperatures, creep and stress relaxation play an important role. Nevertheless, the thrust is generally great enough to increase the fire endurance significantly. In most fire tests of restrained assemblies, the fire endurance is determined by temperature rise of the unexposed surface rather than by structural considerations, even though the steel temperatures often exceed $800^{\circ}C$.

A number of studies have been made on the effects of restraint during fire tests of flexural members (17), (18). Methods of calculating fire resistance of restrained members have not been developed to the same degree as calculations for simply supported or continuous slabs or beams (19), (20). Calculation methods are emperical and are not highly refined. Clearly, additional input into this field would be welcome.

Post-tensioned slabs with unbonded tendons

Use of calculation procedures to determine the fire resistance of post-tensioned concrete beams and slabs which are made with bonded post-tensioned tendons are essentially the same as those for pretensioned prestressed concrete elements. Curved tendons rather than straight or deflected strands introduce only minor differences which do not change the design procedures.

However, with unbonded post-tensioned tendons two differences occur in the design procedures:

1. The stress in unbonded post-tensioned tendons at ultimate load during fire conditions can be derived from the expression:

$$f_{ps\theta}/f_{pu\theta} = f_{ps}/f_{pu}$$

2. For continuous beams or slabs exposed to fire from below, the vaulue of $f_{pu\theta}$ in the negative moment region of continuous draped tendons is the same as that in the positive moment regions.

Example Problem B

To illustrate the use of the procedures described above, assume that a 2-hr fire endurance is required for the 150-mm thick flat plate floor shown on Sheet B1 of the attached calculations. Assume that the floor is one of the typical floors in a multi-story building. Interior panels in the floor can be considered to be restrained against thermal expansion, and thus will readily qualify for a 2-hr fire endurance. The panels between column lines 5 and 6 will have little restraint to thermal expansion in the longitudinal direction, and because the panels are longer than those between column lines 1 and 2, they are probably most critical from a fire endurance standpoint.

Longitudinal tendons are unbonded and are grouped together within the column strips. Transverse tendons are also unbonded but are equally spaced throughout the floor. Supplemental tendons are provided in the end spans. All tendons are 12.5-mm strands with a guaranteed ultimate strength of 1860 MPa. Reinforcing bars are deformed and have a yield stress of 230 MPa, and the 28-day concrete strength is 30 MPa. Slabs are made of carbonate aggregate concrete and the weight of the 150-mm slab can be assumed to be 3.6 kPa. The superimposed dead load is 1.0 kPa and a 2.0 kPs live load is specified.

The first step in the calculations is to determine the retained theoretical moment strength at 2 hr in the positive moment region and in both negative moment regions (over column lines 5 and 6). The moment capacity over line 6 must be transmitted to the columns along line 6, so the maximum value that can be used in the calculations must not exceed that which can be transmitted to the column. It should be assumed that if a fire occurs beneath the floor, a redistribution of moments will occur yielding the negative moment reinforcement. Thus if the span moment is less than the retained moment capacity after redistribution, the fire endurance will be adequate, i.e.,

$$M = M_{n\theta}^{+} + 1/2\,(M_{n\theta 1}^{-} + M_{n\theta 2}^{-})$$

If, however, the span moment is greater than the retained moment capacity at 2 hr, changes must be made in the design. Several options for improving the fire endurance are available. These options include:

1) Increase the concrete cover in the positive moment region.

2) Increase the number of prestressing tendons.

3) Add positive moment reinforcing steel.

4) Add negative moment reinforcing steel.

5) Undercoat the slab with spray-applied insulation.

Of course, there are other solutions such as the use of a thicker slab, the use of lightweight concrete, or the additon of a fire resistive ceiling. Also, combinations of the options listed above can be used. The most appropriate solution depends on in-place cost, architectural acceptability, and perhaps other considerations. For example, to upgrade the fire endurance of an existing floor, options 1 through 4 are not applicable, so either an undercoat or a ceiling might be most appropriate.

Summary of calculations (Sheets B1 through B9 attached)

Option 1 requires that the tendons be raised 5-mm in the positive moment region. Although raising the tendons will result in a 2-hr fire endurance, the behavior and moment capacity at room temperature may be inadequate, so it is likely that two or three additional tendons will have to be added to offset the 17% reduction in tendon eccentricity. Also raising the tendons might create interference with the transverse tendons. These items must be checked before the design is completed.

Option 2 requires addition of two tendons. Behavior at normal temperatures should be checked.

Option 3 requires addition of 69 kg of reinforcing steel per panel in the positive moment region.

Option 4 requires addition of 50.5 kg of reinforcing steel per panel in the negative moment region.

Option 5 requires a 5 mm undercoat of either sprayed mineral fiber or vermiculite cementitious material.

The advantage of utilizing rational design procedures is that a number of options can be analyzed and the best solution based on cost and other considerations can then be selected. Prior to the use of rational design procedures, only one or two solutions were acceptable—increasing the concrete cover or using an undercoat or ceiling. Very often the best solution is the addition of some reinforcing steel which improves not only the fire endurance but also the overall strength and ductility of the floor.

REFERENCES

1. Petterson, O., "Structural Fire Engineering Research Today and Tomorrow", Acta Polytechnica Scandinavica, Ci, Stockholm, 1965.

2. Brockenbrough, R. L., and Johnston, B. G., " Steel Design Manual", U. S. Steel Corp., Pittsburgh, PA.

3. Abrams, M. S., and Cruz, C. R., "The Behavior at High Temperature of Steel Strand for Prestressed Concrete", PCA Research Department Bulletin 134.

4. Gustaferro, A. H., et al, "Fire Resistance of Prestressed Concrete Beams. Study C: Structural Behavior During Fire Tests", PCA Research and Development Bulletin (RD009 01B), Portland Cement Association, 1971.

5. Abrams, M. S., "Compressive Strength of Concrete at Temperatures to 1600°F", Temperature and Concrete, American Concrete Institute Special Publication SP-25, Detroit, MI, 1971.

6. Cruz, C. R., "Elastic Properties of Concrete at High Temperatures", PCA Research Bulletin 191.

7. Cruz, C. R., "Apparatus for Measuring Creep of Concrete at High Temperatures", PCA Research Department Bulletin 225.

8. "Manual of Steel Construction", 7th Ed., American Institute of Steel Construction, New York, NY.

9. Abrams, M. S., "Performance of Concrete Structures Exposed to Fire", 9th National SAMPE Technical Conference, Vol. 9, "Materials and Processes–In Service Performance", Society for the Advancement of Material and Process Engineering, Azusa, CA, 1977.

10. "Guides to Good Practice—FIP/CEB Recommendations for the Design of Reinforced and Prestressed Concrete Structural Members for Fire Resistance", Cement and Concrete Assn., Wexham Springs, England, June 1975.

11. Lin, T. D., and Abrams, M. S., "Temperature Distribution in Rectangular Beams Subjected to Fire", Portland Cement Association, in preparation.

12. Gustaferro, A. H., and Selvaggio, S. L., "Fire Endurance of Simply Supported Prestressed Concrete Slabs", Journal, Prestressed Concrete Institute, V. 12, No. 1, Feb. 1967.

13. Lin, T. D., et al, "Fire Endurance of Continuous Reinforced Concrete Beams", Portland Cement Association Research and Development Bulletin (RD 072.01B), 1981.

14. Ehm, H., and von Postel, R., "Tests of Continuous Reinforced Beams and Slabs Under Fire", Proceedings, Symposium on Fire Resistance of Prestressed Concrete, 1965, FIP, Wexham Springs, England.

15. "Fire Test of a Simple, Statically Indeterminant Beam", Report No. B1-59-22, TNO Institute for Structural Materials and Building Structures, Delft, Holland.

16. Krampf, Lore, "Investigations of the Shear Behavior of Reinforced Concrete Beams Exposed to Fire", Institut fur Baustoffkunde und Stahlbetonbau, Technische Universitat, Braunschweig, Germany. Translation available from C & CA, Wexham Springs, England.

17. Selvaggio, S. L., and Carlson, C. C., "Restraint in Fire Tests of Concrete Floors and Roofs:, ASTM STP 422, American Society for Testing Materials. PCA Research Department Bulletin 220.

18. Issen, L. A., et al, "Fire Tests of Concrete Members: An Improved Method for Estimating Restraint Forces:, Fire test Performance ASTM STP 464, American Society for Testing and Materials, 1970, pp. 153-185.

19. Salse, E. A. B., and Gustaferro, A. H., "Structural Capacity of Concrete Beams During Fires as Affected by Restraint and Continuity", Proceedings, 5th CIB Congress, Paris, France, 1971. Centre Scientific et Technique du Batiment, Paris.

20. Gustaferro, A. H., and Martin, L. D., "Design for Fire Resistance of Precast Prestressed Concrete", Prestressed Concrete Institute, Chicago, IL, 1977.

Notation

a = depth of equivalent rectangular stress block at ultimate load, and is equal to $A_{ps}f_{ps}/0.85\,f_c'b$ or $A_sf_y/0.85\,f_c'b$ (mm)

A_s = area of reinforcing steel (mm^2)

A_{ps} = area of prestressing steel (mm^2)

b = width of compression zone (for use in flexural calculations) (mm)

d = distance between centroid of reinforcement and the extreme compression fiber (mm^2)

f_c' = compressive strength of concrete (MPa)

f_{ps} = stress in prestressing steel in flexural member at ultimate load (MPa)

f_{pu} = ultimate strength of prestressing steel (MPa)

f_{se} = effective stress in prestressing steel after allowance for all prestress loses (MPa)

f_y = yield strength of hot-rolled steel (MPa)

ℓ = span length (m or mm)

M = service load bending moment (kN m)

M_n = theoretical moment strength (kN m)

u = distance from bottom of slab or beam to a point within the member, e.g., the distance from the underside of a slab to the center of a prestressing strand (mm)

w = uniformly distributed load on a flexural member

θ_s = temperature of steel ($^\circ$C)

Subscripts

p = of prestressing steel

s = of reinforcing steel

u = ultimate

θ = as affected by temperature

Superscripts

+ and − indicate positive and negative moment regions

THE CONSULTING ENGINEERS GROUP INC.			SHEET NUMBER
1701 E. Lake Avenue, Glenview, IL 60025 (312) 729-0646		2-hr Fire Endurance Reinforced Concrete	A-1
MADE BY *QHA*	DATE *May 25, '84*	CHECKED BY	JOB NUMBER
		Simple Slab	84201

Example A: Determine the allowable superimposed load for a 2-hr fire endurance for a simply supported 150-mm slab made of siliceous aggregate concrete and reinforced with R16 bars, 125 mm on centers, and centered 25 mm above the bottom of the slab. The span is 4.5 m, $f'_c = 20$ MPa, and $f_y = 230$ MPa.

$$\theta_s = 630^{\circ}C \text{ (Fig. A-1)}; \quad f_{y\theta} = 0.41 \, f_y = 94.3 \text{ MPa (Fig. A-2)}$$

$$A_s = \frac{1000}{125} (200) = 1600 \text{ mm}^2/m; \quad b = 1000 \text{ mm}; \quad d = 125 \text{ mm}$$

$$M_{n\theta} = A_s \, f_{y\theta} \, (d - \frac{a_\theta}{2})/10^6$$

$$a_\theta = \frac{A_s \, f_{y\theta}}{0.85 \, f'_{c\theta} \, b} = \frac{1600 \, (94.3)}{0.85 \, (20)(1000)} = 8.9 \text{ mm}$$

$$M_{n\theta} = 1600 \, (94.3)(125 - 4.5)/10^6 = 18.2 \text{ kN·m} = M$$

$$w = 8M/\ell^2 = 8 \, (18.2)/(4.5)^2 = 7.2 \text{ kPa}$$

slab w = 3.6 kPa, so superimposed w = 7.2 - 3.6 = 3.6 kPa

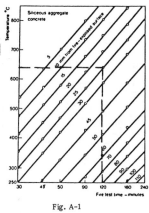

Fig. A-1

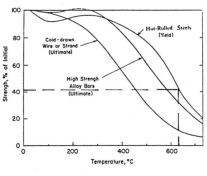

Fig. A-2

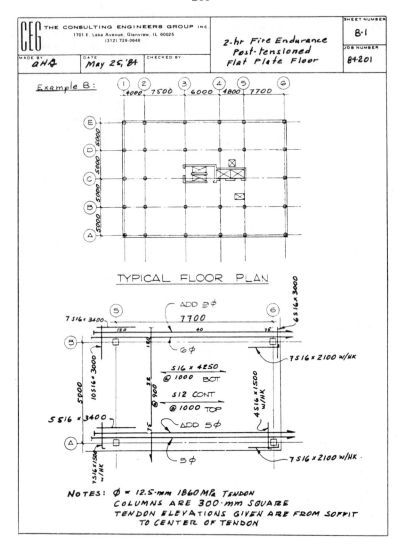

		SHEET NUMBER
CEG THE CONSULTING ENGINEERS GROUP INC. 1701 E. Lake Avenue, Glenview, IL 60025 (312) 729-0646	2-hr Fire Endurance Post-tensioned	B·1
MADE BY *aHQ* DATE *May 25, '84* CHECKED BY	Flat Plate Floor	JOB NUMBER 84-201

Example B:

TYPICAL FLOOR PLAN

NOTES: ϕ = 12.5·mm 1860 MPa TENDON
COLUMNS ARE 300·mm SQUARE
TENDON ELEVATIONS GIVEN ARE FROM SOFFIT
TO CENTER OF TENDON

Column Strip B5 - B6

Loads: DL = 3.6 kPa; SDL = 1.0 kPa; LL = 2.0 Kpa

w = 6.6 kPa

Span Moment $= \dfrac{w\ell^2}{8}$; $\ell = 7700 - 300 = 7400$

$M = \dfrac{5.0\ (6.6)(7.4)^2}{8} = 226$ kN·m

Positive moment capacity @ 2 hr

Reinf. = 15 - 12.5-mm 1860 MPa strands + 5 S16 ($f_y = 230$ MPa)

for strands and bars, u = 40 mm

@ 2 hr, $\theta_s = 460^\circ$C (Fig. B-1);   From Fig. B-2

$f_{pu\theta} = 0.40\ f_{pu} = 744$ MPa

$f_{y\theta} = 0.73\ f_y = 168$ MPa

$A_{ps} = 15\ (101.2) = 1518$ mm^2

$A_s = 5\ (200) = 1000$ mm^2

b   = 5000 mm

d   = 150 - 40 = 110 mm

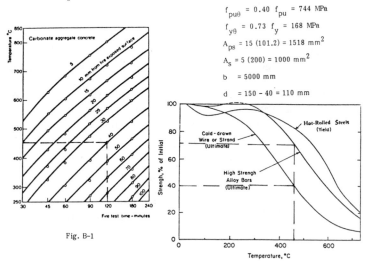

Fig. B-1

Fig. B-2

$f_{ps} = f_{se} + 69 + \dfrac{f'_c}{100\ \dfrac{A_{ps}}{bd}}$

$= 1100 + 69 + \dfrac{30}{(100)\ \dfrac{1518}{5000\ \times\ 110}} = 1277$ MPa

$f_{ps\theta} = \dfrac{f_{ps}}{f_{pu}}\ f_{pu\theta} = \dfrac{1277}{1860}\ (744) = 511$ MPa

$M^+_{n\theta} = A_{ps}\ f_{ps\theta}\ (d - \dfrac{a_\theta}{2})/10^6 + A_s\ f_{y\theta}\ (d - \dfrac{a_\theta}{2})/10^6$

$a_\theta = \dfrac{1518\ (511) + 1000\ (168)}{0.85\ (30)(5000)} = 7.4$ mm

strand $M_{n\theta} = 1518\ (511)(110 - 4)/10^6 = 82.2$ kN·m

rebar $M_{n\theta} = 1000\ (168)(110 - 4)/10^6 = \underline{17.8}$ kN·m

$M_{n\theta} = 100.0$ kN·m

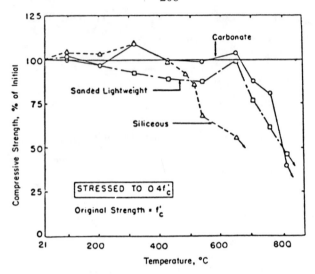

Negative moment capacities - Col. B5

Neglect concrete hotter than 750°C, i.e., bottom 8 mm (Fig. B-1)

$f_{ps\theta}$ = 511 MPa (same as for $M_{n\theta}^{+}$)

Estimate θ_s for top rebars = 150°C (extrapolate from Fig. B-1)

$f_{y\theta}$ = 0.92 (230) = 212 MPa (Fig. B-2); A_s^{-} = 7 (200) + 5 (110) = 1950 mm^2

Assume $f_{c\theta}'$ in a_θ zone = 0.85 f_c' = 25 MPa (Fig. B-3)

$$a_\theta^{-} = \frac{1518 \ (511) \ + \ 1950 \ (212)}{0.85 \ (25)(5000)} = 11.2 \text{ mm; } d_\theta^{-} = 120 - 8 = 112 \text{ mm}$$

strand $M_{n\theta}^{-}$ = 1518 (511)(112 - 6)/10^6 = 82.2 kN·m

rebar $M_{n\theta}^{-}$ = 1950 (212)(112 - 6)/10^6 = __43.8__ kN·m

over Col. B5, $M_{n\theta}^{-}$ = 126.0 kN·m

Col. B6

d_θ^{-} for strands = 75 - 8 = 67 mm

$M_{n\theta}^{-}$ = 1518 (511)(67 - 6)/10^6 + 43.8 = 91.1 kN·m

Check span moment capacity

$$M_{n\theta} = M_{n\theta}^{+} + \frac{M_{n\theta 1}^{-} + M_{n\theta 2}^{-}}{2}$$

$$= 100.0 + \frac{126.0 + 91.1}{2} = 208.5 \text{ kN m} < 226 \text{ \underline{NO GOOD}}$$

Option 1: Raise tendon profile in positive moment region

by 5 mm, so u = 45 mm and d = 105 mm

θ_s = 390°C (Fig. B-1); $f_{pu\theta}$ = 0.56 f_{pu} = 1041 MPa (Fig. B-2)

$f_{y\theta}$ = 0.77 f_y = 177 MPa (Fig. B-2)

f_{ps} = 1100 + 69 + $\dfrac{30}{100 \ \dfrac{1518}{5000 \times 105}}$ = 1272 MPa

$f_{ps\theta}$ = $\dfrac{1272}{1860}$ (1041) = 712 MPa

a_θ^+ = $\dfrac{1518 \ (712) \ + \ 1000 \ (177)}{0.85 \ (30)(5000)}$ = 9.8 mm

strand $M_{n\theta}^+$ = 1518 (712)(105 - 5)/10^6 = 108.1 kN·m

rebar $M_{n\theta}^+$ = 1000 (177)(105 - 5)/10^6 = $\underline{17.7}$ kN·m

$M_{n\theta}^+$ = 125.8 kN·m

Values of $M_{n\theta}^-$ remain the same so

$M_{t\theta}$ = 125.8 + $\dfrac{126.0 \ + \ 91.1}{2}$ = 234 kN·m > 226 OK

> Note: Be sure to check moment capacities and behavior at normal temperatures. Also be sure that transverse steel can be installed without interference. Because the tendon eccentricity is reduced from 35 mm to 30 mm, it is likely that two or three additional tendons will be needed.

Option 2: Add tendons to end span

Assume $M_{n\theta}^+$ and $M_{n\theta}^-$ are increased equally by

$\dfrac{226 \ - \ 208.5}{2}$ = 8.8 kN·m

strand $M_{n\theta}^+$ = 82.2 kN·m

Approx. No. of strands = $\dfrac{82.2 \ + \ 8.8}{82.2}$ (15) = 17

A_{ps} = 17 (101.2) = 1720 mm^2; f_{ps} = 1266 kN·m

$f_{ps\theta}$ = 506 MPa; a_θ^+ = 8.1 mm

$M_{n\theta}^+$ = 110 kN·m

a_θ^- = 12.1 mm; @ Col. B5, $M_{n\theta}^-$ = 126 kN·m

@ Col. B6, $M_{n\theta}^-$ = 97 kN·m

$M_{t\theta}$ = 110 + $\dfrac{136 \ + \ 97}{2}$ = 226.5 kN·m > 226 OK

> Note: Be sure to check behavior at normal temperatures.

Option 3: Add positive moment rebars

Increase $M_{n\theta}^+$ by $226 - 208.5 = 17.5$ kN·m

Estimate A_s^+ (rebars) needed

Before adding rebars, $M_{n\theta}^+ = 100$ kN·m and $a_\theta^+ = 7.4$ mm

New $M_{n\theta}^+ = 100 + 17.5 = 117.5$ kN·m and $a_\theta^+ = \frac{117.5}{100}$ $(7.4) = 8.7$ mm

Strand $M_{n\theta}^+ = 1518 \ (511)(110 - 5)/10^6 = 81.4$ kN·m

rebar $A_s^+ = \frac{(117.5 - 81.4) \ 10^6}{168 \ (110 - 5)} = 2046$ mm^2

Try 11 S 16; $A_s = 2200$ mm^2

$a_\theta^+ = \frac{1518 \ (511) + 2200 \ (168)}{0.85 \ (30)(5000)} = 9.0$ mm

rebar $M_{n\theta}^+ = 2200 \ (168)(110 - 5)/10^6 = 38.8$ kN·m

$M_{n\theta}^+ = 81.4 + 38.8 = 120.2$ kN·m > 117.5 OK

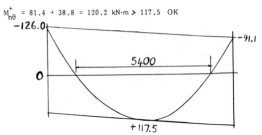

Length of rebars $= 5400 + 2 \ \ell_d = 6000$

Wt. of M^+ bars $= 11 \ (6.0)(1.55) = 102$ kg

Orig. wt. of M^+ bar $= 5 \ (4.25)(1.55) = \underline{33}$ kg

Added rebar wt $= 69$ kg

Option 4: Add negative moment rebars over Col. B5

$M_{n\theta}^-$ must be increased by $2 \ (226 - 208.5) = 35.0$ kN·m

$M_{n\theta}^-$ must $= 126.0 + 35.0 = 161.0$ kN·m

$a_\theta^- = \frac{161.0}{126.0}$ $(11.2) = 14.3$ mm

strand $M_{n\theta}^- = 1518 \ (511)(112 - 8)/10^6 = 80.6$ kN·m

rebar $A_s^- = \frac{(161.0 - 80.6) \ 10^6}{212 \ (112 - 8)} = 3646$ mm^2

Try 10 S20 + 5 S12, $A_s^- = 3650$ mm^2

$a_\theta^- = 14.6$ 14.3

rebar $M_{n\theta}^- = 3650 \, (212)(112 - 8)/10^6 = 80.5$ kN·m

$M_{n\theta}^- = 80.6 + 80.5 = 161.1$ kN·m > 161.0 OK

Contribution of continuous S12 bars $= \dfrac{550}{3650} \, (80.5) = 12.1$ kN·m

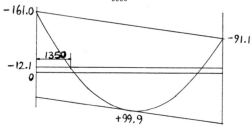

Length of added rebars = 2 (1.35) + 0.3 + 2 (0.3) = 3.6 m

Wt. of S20 bars	= 10 (3.6)(2.43)	=	87.5 kg
Wt. of Orig. S16 bars	= 7 (3.4)(1.55)	=	37.0 kg
	Added rebar wt.	=	50.5 kg

Option 5: Undercoat slab with spray-applied insulation

From Option 1, 5mm added cover is adequate. Thus an added equivalent concrete cover thickness of 5mm will be adequate. From Fig. B-4, less than 5 mm of sprayed mineral fiber or vermiculite cementitious material will be adequate. The minimum practical thickness is about 5 mm. So use 5 mm of SMF or VCM.

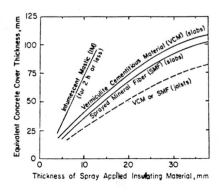

Fig. B-4

FIRE TESTS ON RIBBED CONCRETE FLOORS

LAWSON, ROBERT MARK

Construction Industry Research and Information Association
London SW1P 3AU

ABSTRACT

This paper describes the results of large-scale fire tests on various
forms of ribbed concrete floors. The test floors were representative
sections of real structures incorporating the moment-continuity which was
considered to be a major beneficial effect in fire. The tests showed that
waffle slabs, comprising a criss-cross of ribs, could resist load for over
2 hours without measures to control spalling of the concrete in fire. A
test on a composite deck slab survived 1½ hours, which is satisfactory for
most commercial buildings.

INTRODUCTION

When thin reinforced concrete sections with normal weight siliceous
aggregate are exposed to fire, spalling of the concrete cover to the
reinforcement can occur. The reinforcing bars are then exposed directly to
heat, leading to a loss in strength and stiffness of the section. In
design to the former UK Code of Practice for reinforced concrete, CP110 :
1972 [1] supplementary reinforcement was required for concrete covers of
40 mm or greater to help maintain this concrete around the bars. The
implication of this for design was that many forms of ribbed concrete
floors would require supplementary reinforcement for fire resistances of 2
hours and above. The problem was compounded for waffle slabs which have
overlapping bars in the ribs and hence variable cover.

The real behaviour of continuous beam or slab elements is that after
spalling has taken place, considerable reserve of safety exists because of
the redistribution of moment from the heat affected zone in mid span to the
support zone. This is often reflected in the observations of fire-damaged
structures which perform well, even after serious spalling. This is the
basis of the fire engineering method of design presented in the Institution
of Structural Engineers Report [2].

The aim of the research set up by CIRIA was to show that continuous
ribbed floors may be designed for fire resistances of 2 hours without
requiring supplementary reinforcement. This centred around three large-
scale fire tests performed between November 1984 and March 1985. The
results would confirm the revised fire resistance tables in BS8110 [3], the
updated Code of Practice for reinforced concrete (then in draft).

A subsidiary series of fire tests is currently underway to examine the spalling characteristics of different concretes in fire, following on from the desk study for CIRIA by H.L.Malhotra [4]. Part of this work involves testing slabs where the concrete cover to the bars in the rib has been repaired by various methods.

A complementary project was carried out in 1984 to assess the fire resistance of a related form of ribbed floor where steel decking is used as permanent formwork to the concrete slab. The floor with its support steelwork would be fire-tested to show that, despite possible debonding of the deck, the floor could achieve a fire resistance of 1½ hours.

FIRE TESTS ON WAFFLE SLABS

The test slabs were designed to be representative of the 'column-strip' of a real two-way spanning waffle slab floor. Each slab incorporated short columns and a cantilever section which was loaded by a jack to develop moment-continuity between spans. The pair of small edge columns stabilized the slab. The slab size of 6.8 m length x 2.9 m width was the largest possible that would fit within the test furnace at the Fire Insurers' Research and Testing Organization (FIRTO). The slab span of 5.4 m comprised six waffle units at 900 mm spacing. The mould size of 225 mm depth with a topping thickness of 80 mm meant that the slab thickness was a practical minimum of 305 mm.

The slabs were designed by the elastic method in CP110 [1] to resist an imposed load of 5.5 kN/mm^2. Because of the concentration of moment in the 'column-strip' of a real slab, the actual load on the test slab was increased in intensity in order to develop the same hogging and sagging moments, as in the real slab. The details of the reinforcement in the slab are shown in Figure 1 and the test slab under load is shown in Figure 2.

A jack force at the end of the cantilever was applied to correspond initially to the omitted load on the adjacent span and would be increased during the fire to model the build up of hogging moment as in Figure 3. This also imposes an extension of the hogging zone into mid span. The test details were:

Test 1 : Normal weight (gravel aggregate) concrete (NWC). Lower pair of bottom bars in rib in span direction at 35 mm cover.

Test 2 : NWC. Lower pair of bottom bars at 20 mm cover.

Test 3 : Lightweight (Lytag aggregate) concrete (LWC). Lower pair of bottom bars at 20 mm cover.

All the slabs were cast at the University of Aston about 8 months prior to testing and had reached a moisture content of around 3% (by weight) for NWC and 5% for LWC. These were considered to be representative of a recently occupied building. The measured 28-day cube strengths of the concrete were between 38 and 45 N/mm^2, well in excess of the required grade 30. The measured yield strength of the 16 mm diameter bars in the ribs was 534 N/mm^2 (nominal strength 485 N/mm^2).

215

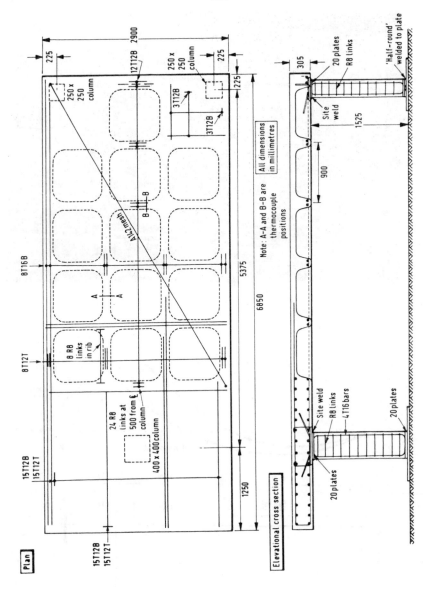

Figure 1   Reinforcement in test waffle slab and columns

Figure 2   View of slab in furnace prior to fire test

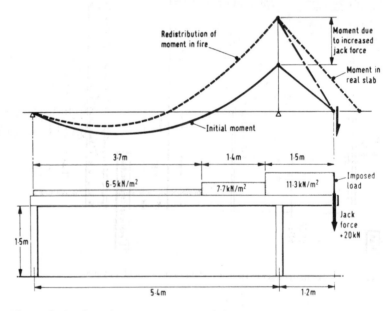

Figure 3   Loads and moments on test slab

RESULTS OF FIRE TESTS

Some upward movement of the cantilever occurred during the early stages of the fire arising from temperature induced curvature. In test 1 there was only limited upward travel of the jack and its force therefore increased to depress the cantilever. This effect is shown in Figure 4. However, after two hours of the fire this effect diminished and the structural weakening of the slab was such that the increase in deflection in mid span was more rapid. At the end of the test, which was stopped after 4 hours, the mid span deflection had not reached the limiting value of span/30 specified in BS476 part 8 [5]. The total load at the end of the cantilever was 70 kN (jack force + 20 kN), an increase of some 40% over the initial condition.

In Test 1, spalling of the concrete cover had begun after about 30 minutes and large sections had become detached, particularly from the transverse ribs where the cover was greatest. Because of this the rise of temperature of the bars in the ribs was relatively rapid (Figure 5). In Test 2, little spalling occurred until after 2 hours, presumably because of the smaller cover, and this slab also survived 4 hours.

For Test 3, it was decided to maintain the jack force at zero for as long as possible to assess the performance of the lightweight slab as a simply-supported element. Consequently, deflections were much greater than in Test 1 but nevertheless the jack load was not reapplied until after 2 hours. Temperatures in the reinforcement and in the concrete were some 20% less than recorded in Test 2 (Figure 6), which explains the slab's better performance.

The double-curvature of test slab 1 is illustrated in Figure 7 as it is lifted out of the furnace. The spalling of the concrete from the ribs can be seen, along with the cracking caused by the rotation of the edge columns. This suggests that good detailing of the reinforcement especially anchorage of the bars, is necessary for good performance in fire.

The residual yield strength of the bottom reinforcements in Test 1 was measured as 305 N/mm^2, which show that the slab would be unserviceable after such an intensity of fire. This result agrees with the data given by Holmes et al [6] for cold worked bars having reached a peak temperature of roughly 800 C.

From the fire tests it can be concluded that continuous ribbed slabs perform adequately in fire for fire resistance periods of at least 2 hours. Lightweight concrete slabs can resist load for this period as simply-supported elements. The theoretical capacities of the slabs were calculated using the fire engineering method [2] which proved to be conservative. The results of this research are described fully in CIRIA Report 107 [7].

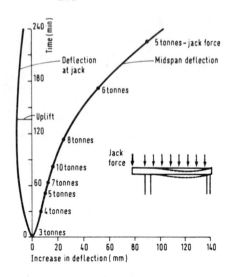

Figure 4   Deflection of waffle slab during fire test (jack force
acting as a restraint)

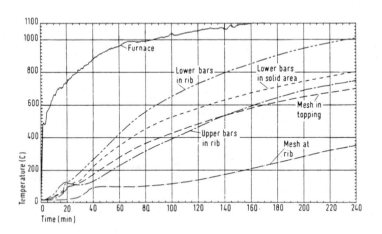

Figure 5   Temperature rise during Test 1

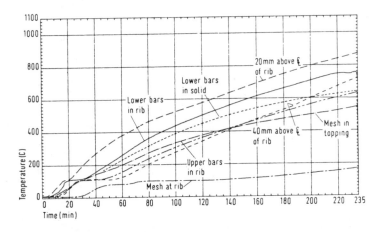

Figure 6   Temperature rise during Test 3

Figure 7   Test slab being lifted out of furnace

SMALL-SCALE TESTS ON WAFFLE SLABS

The large-scale tests had shown the influence of continuity but more information was needed on the spalling characteristics of concrete in thin ribs including the effect of permanent formwork systems and concrete repair methods. Many of these factors had been previously highlighted [4]. Using the 225 mm deep waffle slab as a basic section shape a limited series of tests were initiated to fit in a 1.5 m x 1.0 m slab furnace. The test parameters were:

Concrete aggregates : Gravel, limestone, granite and LWC.

Section shape   : Waffle rib and rectangular rib of 120 mm width.

Reinforcement   : Two 16 mm bars or one 25 mm bar in ribs at 30 mm cover in longitudinal rib.

Additional bars   : Shear links and supplementary mesh.

Concrete repairs   : Epoxy resin, acrylic, SBR (latex) and cementitious repairs to concrete cover.

Surface protection   : GRC permanent formwork moulds 5 mm and 10 mm thick.

The test series is still in progress. Nevertheless those tests which have been carried out indicate that slabs with GRC permanent formwork can survive up to 4 hours without spalling and offer 10 to 20% reduction in temperature relative to plain gravel concrete of the same cover. Similarly, the rise of temperature of limestone aggregate concrete was found to be some 10% less than gravel aggregate. These reductions are significant as they may contribute to a 20% increase in the yield strength of the steel at these elevated temperatures [2]. The beneficial effect of LWC has already been noted. Both shear links and supplementary mesh are successful in controlling spalling.

FIRE TEST ON COMPOSITE SLAB

The influence of structural continuity is also a major factor in the fire resistance of composite deck slabs, where the mesh in the concrete topping provides the necessary moment resistance at the slab supports. The performance of this form of ribbed slab is more affected by heat than in reinforced concrete because the deck is exposed and can debond.

In order to justify the design of composite slabs with nominal A142 mesh reinforcement for a fire resistance of 1 hour, which is specified for most commercial buildings, a large-scale fire test was devised by CIRIA. The special features of this test were the two complete bays of 3 m span and the realistic proportions of the beams, columns and their connections (Figure 8). In order to represent the behaviour of a composite beam of 8 m span within the constraints of the furnace, a special sliding detail was devised which would allow the beam to pull-in during the fire.

The deck shape chosen was representative of most trapezoidal profiles and the topping was in lightweight concrete giving a total slab depth of

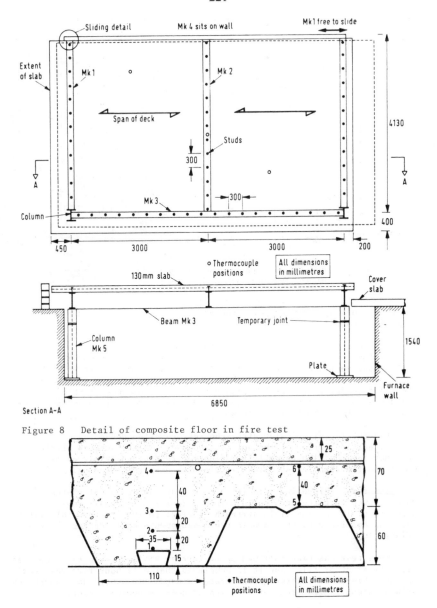

Figure 8   Detail of composite floor in fire test

Figure 9   Section through composite slab

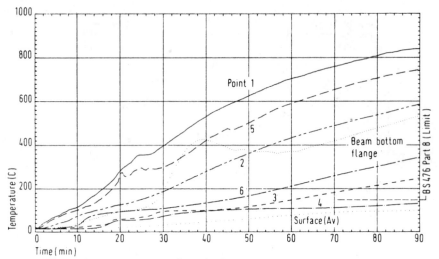

Figure 10   Temperature rise within composite slab (see Figure 9
for thermocouple points)

130 mm (Figure 9). The support steelwork was protected to achieve a
nominal 1½ hour fire resistance.

In the fire test the slab deflected at an almost constant rate of
0.5 mm/minute with little debonding until after 1 hour, at which stage
deflections increased more rapidly. The test was stopped after 1½ hours
when the absolute deflection of the slab approached span/20 or 150 mm.
The relative deflection of the slab between the beam ends at this stage
was about 100 mm or span/30. The slab and its support frame continued to
resist load for a further 24 hours.

Examination of the slab after the test revealed debonding over a
significant proportion of the soffit. However, there had been little
tendency for the slab to pull-in indicating that cetenary forces were
small and the slab resisted load in flexure, despite debonding.

The fire engineering capacity of the slab was again calculated [8]
which showed that the deck would have to provide a residual tensile
strength of 10% of its yield strength in order to justify this calculation.
This was confirmed by the temperature measurements taken during the test
(Figure 10).

This test has been followed up by further smaller-scale tests on slabs
with different profiles and it is concluded that a 1½ hour fire resistance
can be achieved if lightweight concrete is used. In other cases additional
reinforcement is required for fire resistances exceeding 1 hour.

CONCLUSIONS

The large-scale tests on reinforced concrete waffle slabs have shown that they may be safely designed for fire resistances of 2 hours without measures to control spalling, provided that the reinforcement is well detailed to develop structural continuity between adjacent spans. Lightweight concrete slabs can survive 2 hours as simply-supported elements. These conclusions also apply to trough and other forms of ribbed slab. Composite deck slabs can be safely designed for $1\frac{1}{2}$ hours fire resistance with mesh reinforcement if lightweight concrete is used.

REFERENCES

1. British Standards Institution. The structural use of concrete, CP110 : 1972.

2. Institution of Structural Engineers. Design and detailing of concrete structures for fire resistance, 1978.

3. British Standards Institution. The structural use of concrete, BS8110 : 1985.

4. Malhotra, H.L., Spalling of concrete in fires. CIRIA Technical Note 118, 1984.

5. British Standards Institution. Fire tests on building materials and structures. BS476 part 8. Test methods and criteria for the fire resistance of elements of building, 1972.

6. Holmes, M.; Anchor, R.D.; Cooke, G.M.E.; Crook, R.N., The effects of elevated temperatures on the strength properties of reinforcing and prestressing steel. The Structural Engineer (Part B), 60 (1), March 1982, 7.

7. Lawson, R.M., Fire resistance of ribbed concrete slabs. CIRIA Report 107, 1985.

8. Newman, G.M.; Walker, H.B., Design recommendations for composite floors and beams Section 2 : Fire resistance. Constrado, 1983.

FIRE RESISTANT STEEL DESIGN - THE FUTURE CHALLENGE

J.T.ROBINSON and D.J.LATHAM

British Steel Corporation

J.T. Robinson is Section Manager Market Development at BSC
Sections

D.J. Latham is a Principal Research Officer at BSC Swinden
Laboratories

ABSTRACT

    Current research is throwing new light on the behaviour
of steel structures in fire. The effects of load and section
dimensions on the fire resistance of fully exposed steel
members are defined.
    Results of standard fire tests on partially exposed
members show that unprotected steelwork can be designed to
withstand high temperatures and that significant levels of
fire resistance can be achieved without protection by
partially shielding structural elements with walls and floors.
Further work is necessary to realise the full potential of the
findings but over the latter half of this decade increased
understanding of high temperature behaviour could enable
assessment of stability in fire to become an integral part of
the design process.

INTRODUCTION

    In the UK we can look back over five years and see a
steadily rising trend in the use of steel for building frames,
based largely on reduced raw material cost and improved
productivity. It has been a time of radical change in the
economic structure of the industry driven by the steelwork
producers and fabricators. While costs of production and
fabrication will continue to fall in the future - with the
increasing use of advanced techniques, such as continuous
casting of liquid steel and computerised fabrication processes
- the major challenge and opportunity of the latter half of
the decade will lie in the hands of the engineer.

    Since steel was first used for building frames about 100
years ago most development effort has been concentrated on
reducing the steel weight and, while great strides have been
made, there will be less scope for further major economies in

this area in the future. To a large extent the same can be said of fabrication.

A survey carried out in the UK in 1982 indicated that the average cost of fire protection represented about 30% of the total cost of a multi-storey steel frame. Raw steel and fabrication each accounted for about 30% with the remaining 10% being taken up by supply and erection. If the cost effectiveness of steel framed construction is to be improved then we must first seek to minimise the high cost areas.

In the future research and development will, of course, continue to be carried out into many aspects of structural engineering, but the greatest opportunity for improving cost effectiveness lies in new approaches to the problem of fire.

The heart of the problem lies in the fact that all materials suffer a reduction in strength at high temperatures and steel, having a relatively high thermal conductivity compared to other structural materials such as concrete, is considered to be particularly prone to weakening under fire conditions.

Engineers have traditionally designed building frames to be the most efficient structures they can devise utilising the ambient temperature properties of the material. Insulation is then applied to steel structures to ensure that the material properties used in the design (ambient temperature) remain valid when the environment temperature rises in fire conditions.

This approach raises two questions. Is it possible to design in steel using high temperature properties, rather than ambient temperature properties, and thus reduce the need for insulation ? If it is possible, is it economic ?

Much work remains to be done before these questions can be answered in detail and the limits defined, but the experimental data presented in this paper suggest that the answer to both questions is "yes" and that a "Fire Resistant Design" approach could be developed which would minimise the need for protective insulation.

TRADITIONAL CONCEPTS

The need to provide structural stability in fire is self evident. The traditional approach is a qualitative one based on fire loads anticipated for broad categories of building eg. "storage", "offices" etc. The fire environment in each case is expressed as the required time of survival in a standard fire (BS 476, ISO 834 etc) and stability has generally been measured on fully exposed single elements subject to the maximum permissible load to simulate the worst possible case.

Steel columns are axially loaded in the standard fire test and, being exposed on all four sides, are uniformly heated. When tested at the maximum permitted load, which in the UK is defined by BS449 and is approximately 60% of the ambient temperature yield stress, failure occurs when the steel member reaches a temperature of about 550°C. This finding coincides with the results of elevated temperature tensile tests carried out in laboratories on small, axially loaded and uniformly heated, samples of strucural steel (from which the well known graphs of strength -v- temperature are derived). The load bearing capacity in these tests is also reached at around 550°C when the load is equivalent to about 60% of ambient temperature yield stress.

As a result 550°C (with slight variations in other countries) has become widely known as the "Critical Temperature" of steel. Although appealing in it's simplicity the concept of a unique "Critical Temperature" is unsound. A member will fail at 550°C only under the particular conditions of maximum permissible axial load and uniform temperature distribution throughout it's cross section - conditions that rarely occur in practice. Temperature gradients within a member can profoundly influence its performance as the experimental data, described later in this paper, show.

MODERN CONCEPTS

If we wish to refine our knowledge and quantify the behaviour of structures in fire we can approach the problem in two ways, either from the point of view of the fire - what temperature will be generated in a fire and what temperature will the structure reach ? - or from the point of view of the structure - how will the structure perform and what temperature will it withstand ?

A great deal of extremely valuable work has been carried out, largely in the last 20 years, on the first approach - quantifying the fire. Theoretical models now exist which enable the combustion temperature generated in a natural fire to be predicted with reasonable accuracy and from this the temperature of the structural elements.
Data provided by the models can, if necessary, be translated into standard fire test terms by means of the "time equivalent" concept which enables calculated structure temperatures to be converted to traditional fire resistance times.

However in translating calculated structure temperatures into structural performance most of the theoretical treatments adopt the critical temperature concept and assume uniform temperature distribution within the steel members.

It is the second approach, that of quantifying

structural performance, and the temperature that the structure
can withstand, towards which research effort now needs to be
directed.

The experimental work described in this paper represents
a contribution to this research.

EXPERIMENTAL PROCEDURES

The investigations described in this paper were
conducted by BSC Swinden Laboratories. The experimental data
were derived from tests carried out under the standard time –
temperature conditions defined in BS476 pt 8. Beam tests were
undertaken at the Warrington Research Centre and column tests
at the FIRTO fire test facility at Borehamwood.

Standard, rather than natural, fire conditions were
chosen for this phase of the work for two reasons. Firstly to
ensure compatibility with current building regulation
requirements and to enable the results to be compared directly
with the large body of existing data on protected elements
which has been gathered over many years. Secondly structural
behaviour is independent of fire conditions, it depends only
on the steel temperature. The fire severity influences the
survival time but the purpose of this phase of investigation
was to determine failure temperatures. For this reason a
controllable but growing fire environment throughout the test
was necessary.

The beam tests described were conducted in a gas fired
furnace heated so that the atmosphere temperature followed the
ISO time-temperature curve. The span was 4.5 metres with at
least 4.0 metres being exposed to the flames. The furnace roof
comprised strucural quality concrete slabs in direct contact
with the upper flange of the test beam to simulate normal
pre-cast floor constuction. Loads were in accordance with
BS449 which stipulates a maximum permissible design stress of
165 N/sq.mm for BS4360 grade 43 steel and 230N/sq.mm for grade
50 steel. Loads were applied by hydraulic jacks to the
concrete cover slabs and the "failure" criterion was taken as
a vertical deflection of span/30 ie. 150mm deflection for a
4.5 metre span.

Columns, of 3 metre height, were tested in a vertical
gas fired furnace under ISO 834 conditions. Axial loads were
again in accordance with BS449 and were calculated, following
normal test station procedure, on the basis of an effective
length of 0.7.

FULLY EXPOSED ELEMENTS

It has long been recognised that the heating rate, and
therefore the fire resistance, of a steel section depends on
its dimensions or mass. Light sections heat up more rapidly

than heavy sections. For this reason UK building regulations have differentiated between heavy sections (columns over 45kg/m, beams over 30kg/m) and lighter sections. More recently the relationship between dimensions and heating rate has been expressed with greater accuracy by the Hp/A ratio or "Thermal Response Factor". Hp/A is the ratio of the length of the section perimeter exposed to the fire, Hp, to the cross sectional area, A.

It has also long been recognised that the stress in a member influences it's failure temperature. Laboratory elevated temperature tensile tests show that yielding (onset of plasticity) occurs at progressively lower levels of stress as the temperature is raised. In structural terms – the lower the stress in a member the higher the temperature at which excess deformation, and thus failure, occurs.

One of the first objectives of the research programme was to quantify the relationship between section dimensions, stress and fire resistance.

Beams

A total of 28 tests have been carried out on unprotected steel beams covering a range of sizes and test loads. These tests allowed the heating rate, in the standard fire, of a range of beam sizes to be determined. In parallel, a mathematical model has been developed by BSC Swinden Laboratories to predict the changing temperature profiles in beams subject to the standard fire test using the physical laws governing radiation and convection and the thermal properties of steel.

Figure 1 shows the relationship between section dimensions, Hp/A, and time for the lower flange to reach 550, 650 and 750°C in the standard fire test. The relationship predicted by the mathematical model is superimposed on the experimental data and shows that the model can be used

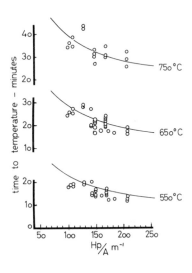

FIGURE 1 Heating rates of unprotected beams

with confidence to reflect observed behaviour. The scatter of results is largely due to manufacturing variance in steel properties and the fact that the test furnace rarely follows the ISO time-temperature curve exactly.

Figure 2 shows the relationship between lower flange temperature and stress. The limiting deflection of L/30 ocurred in beams subject to the maximum permissible stress of BS449 (165N/sq.mm) when the lower flange reached a temperature between 630 – 670°C. At lower stresses the limiting temperature of the lower flange is increased.

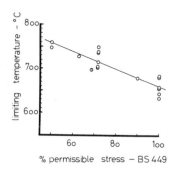

FIGURE 2 Effect of stress on limiting temperature

The two curves shown in figures 1 & 2 can be combined, as in figure 3, to show the relationship between section dimensions, stress and fire resistance for unprotected beams exposed to the standard fire on three sides.

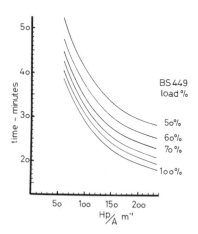

FIGURE 3 The effect of load and section size on the fire rating of unprotected beams

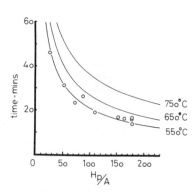

FIGURE 4 Heating rates of unprotected columns

Columns

The performance of bare columns has been examined in a similar way. Figure 4 shows the heating rate curves predicted by the model superimposed on test data representing time to reach 550°C in the standard fire which has been gathered from a variety of sources. Many of the data points are for unloaded indicative tests and although the data available at this time for loaded columns is not as comprehensive as for beams it is clear that 1/2 hour fire rating is limited to fully loaded bare columns having a thermal response factor (Hp/A) less than 50.

TEMPERATURE GRADIENTS

In most model treatments of the behaviour of steel in fire it is assumed, partly through lack of data and partly in order to simplify a complex subject, that members are heated uniformly. This is clearly a very broad assumption - we know from the test results described above that unprotected beams, which are exposed to heat on three sides, exhibit marked temperature gradients. Typically failure occurs when the lower flange temperature is around 650°C and the upper flange is about 450°C.

At first sight it appears that the average temperature at failure is about 550°C and that the "critical temperature" concept still applies to unevenly heated members. But this is simply coincidence.

Theoretically, temperature gradients within a member should improve performance - the principle being that in an unevenly heated member, where one flange is heated faster than the other, the hot flange will reach its elastic limit while the cool flange is still acting elastically. Instead of failure occuring at this point the hot flange will yield plastically and transfer load to the cool flange. The member will retain stability until the load transferred into the cool flange is sufficiently high to cause it too to yield plastically. The precise hot flange temperature at which an unevenly heated member fails will depend on the severity of its temperature gradient and the level of applied stress.

If we accept this concept, that failure temperature is a variable, then a new failure criterion must be introduced to replace the fixed assumption of "critical temperature".

LIMITING TEMPERATURE

The forthcoming structural steelwork code, BS 5950 pt8 will introduce the concept of "Limiting Temperature". This is the temperature that the hottest part of a section must reach at any given load in order to cause failure of the member. It is a much more realistic concept that recognises that the performance of a member will vary according to its load and the temperature profile that develops in fire conditions. The standard will give values of limiting temperature for a range of exposure conditions and load factors.

CURRENT RESEARCH

The objective of the current research phase is to determine how limiting temperature is affected by temperature gradients and how this might be put into practice. The work is still in progress but results to date confirm that fire resistance is improved.

Columns

Column tests have been carried out using the 203x203x52 size, which is the normal section for standard fire resistance tests on protection materials.

When fully exposed without protection, and therefore heated uniformly, this section has a fire resistance rating under maximum BS 449 load of around 15 minutes with a limiting temperature of 550 °C. When the exposed perimeter of the section was reduced by laying blockwork between the flanges, leaving the outer faces of the flanges bare as illustrated in figure 5, the limiting temperature (of the hot flange) at maximum load, was raised to 625°C by transference of load from the hot flanges to the cool web, and the fire resistance time was increased to 36 minutes.

FIGURE 5 Column partially shielded with blockwork. Two flanges exposed.

A second test using 60% of the maximum load permitted by BS449 reached a limiting temperature of 700°C and a fire resistance time of 38 minutes.

Columns embedded in partitions or outer walls of buildings (figure 6) experience even stronger temperature gradients under fire conditions and higher limiting temperatures and fire resistance times are achieved. Tests have been carried out on 203x203x52 columns as illustrated in figure 6. In all cases loads were applied axially and care was taken to ensure that loads were carried solely by the columns and not by the lightweight concrete blockwork walls.

233

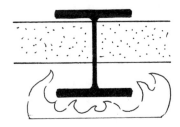 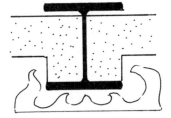

FIGURE 6a   FIGURE 6b
Columns partially shielded by blockwork walls. One flange
exposed.

An open web column (figure 6a) tested at a load slightly
above the maximum BS 449 design load (113%) exhibited a
limiting temperature of 736°C and a fire resistance time of 30
minutes. A second test at 57% of maximum load gave a limiting
temperature of 1002°C and a time of 103 minutes.

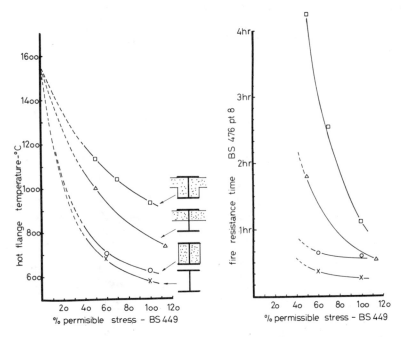

FIGURE 7a   FIGURE 7b
The effect of load on limiting temperature and fire rating for
partially exposed columns.

A minor modification to the test design, acomplished by extending the blockwork to the inside of the exposed flange (figure 6b), resulted in a significant improvement in both limiting temperature and survival time. At 100% of permitted load the limiting temperature reached 934° before failure at 63 minutes, at 70% load the corresponding figures were 1041°C and 153 minutes, and at 40% load the column was showing no sign of failure after 4 hours so the load was increased to 50% and failure subsequently ocurred at 1153°C at 255 minutes. These results are shown graphically in figure 7.

Figure 7a incorporates a hypothetical origin point which is the melting temperature of steel. It seems reasonable to assume that a member subject to zero stress, not even self weight, would not deform until the melting point was reached (at which point we could expect a weightless column or beam to shrink to a sphere under the influence of surface tension).

Clearly uneven heating of columns under axial load can substantially improve their fire resistance but the full practical significance of this finding will require further analysis. Temperature gradients across a section will,for instance, induce thermal bowing and slender columns may require special consideration. Also, tests to date have not considered the effect of bending as a consequence of expansion from a connecting beam.

Beams

The effect of temperature gradients and partial exposure of beams has also been investigated in the research programme.
When pre-cast concrete slabs are used for floor construction they are normally positioned on the top flange of the support beams. They can, however be placed under the top flange on shelf angles fixed to the beam web as shown in figure 8. In this form the top flange, which is normally covered by a screed, and the top portion of the web are not exposed to the fire environment and strong temperature gradients are developed.

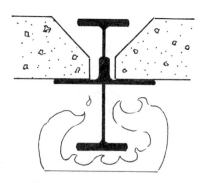

FIGURE 8 Unprotected beam embedded in pre-cast concrete floor slabs.

Beam tests in this phase of the work have been carried out mainly using sections of 406x178x54 size in grade 43 mild steel, in all cases the beams were simply supported without restraint. Tests have been carried out using a range of applied loads and slab depths have been selected to enable 50% and 75% of the web depth to remain exposed. The maximum permitted design stress in the structural code BS449 is 165N/sq.mm and the limiting temperature in this case is the temperature of the lower flange when the beam assumes a deflection of L/30.

With 75% of the web exposed and maximum applied load the limiting temperature was found to be 733°C and the fire resistance time 29 minutes. At 60% load the corresponding figures were 797°C and 43 minutes.

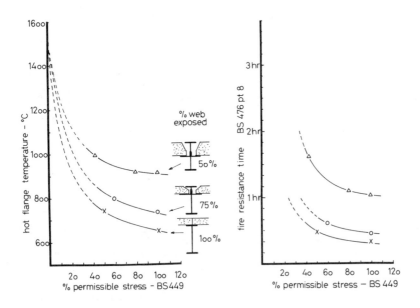

FIGURE 9a   FIGURE 9b
The effect of load on limiting temperature and fire rating for partially exposed beams.

When the slab thickness was increased and exposure reduced to 50% of the web depth the limiting temperature at maximum load increased to 915°C and the fire resistance time to 67 minutes. At 80% load the figures were 914° and 70 minutes, and at 43% load 992°C and 94 minutes. These results are shown graphically in figure 9.

The develpoment work described in this paper represents only one of a number of investigations carried out by BSC into the performance of unprotected steelwork in fire. Other studies, some of which are joint projects with the BRE Fire Researchh Station , aim to 1) obtain a better understanding of the basic material properties at high temperature. 2) examine the effects on the behaviour of steel frameworks of connections and interaction between individual steel elements. 3) Use mathematical modelling to predict the performance of elements in fire. 4) Measure the heating characteristics of unprotected steel members in real fires and relate these to standard furnace tests.

The results will provide an opportunity to extend the utilisation of unprotected steelwork in buildings by the development of appropriate design guidelines and through application of fire engineering principles.

BENEFITS

The cost of fire protection of multi-storey frames can account for up to £10/sq.m of floor area for fire ratings between one and two hours. The ability to design frames without costly protection using existing elements of the structure, such as walls and floors, to provide the necessary fire resistance is likely to reduce this cost directly especially in the lower fire rated classes of buildings. In addition it will improve scheduling and cut construction time.

The major benefits however would be gained indirectly in the areas of cladding and services - which together can account for up to three-quarters of the cost of modern buildings.

Designers have the opportunity to reduce beam downstands and produce slimmer floors at no extra charge by offsetting the cost of shelf-angle beam construction against the cost of fire protection

Slimmer floors can lead to cheaper, simpler service runs, reduced heating demand, and easier upgrading of services at a later date.

Reduced floor to floor height offers the potential of reduced cladding cost, particularly when brickwork rather than pre-cast panelling is used, could allow an additional storey when height restrictions apply, and gives greater flexibility in elevational treatments especially for in-fill sites.

REPAIR OF FIRE DAMAGED STRUCTURES

C.D. JONES

Freeman Fox (Wales) Ltd.
Consulting Engineers
26/27 St. Mary Street
Cardiff CF1 2PD

SUMMARY

This paper attempts to summarise and update the current practices for assessment and repair of fire damaged structures, with particular reference to reinforced concrete. There are continuing developments in the fields of materials, structural forms and design which make the need for careful investigation and repair more complex than in the past and there is therefore a continuing need to review repair methods

INTRODUCTION

There are a number of publications which give advice on the assessment and repair of fire damaged structures (1-4). When we think of repairs to fire damaged structures we usually think first, and perhaps last, about reinforced concrete repairs and this subject is covered by the Concrete Society Technical Report No.15 (1) "Assessment of fire-damaged structures and repair by Gunite". That publication is currently being revised by the Fire Resistance Committee of the Concrete Society and this paper outlines some of the major revisions. It is expected that the revised report will be available in mid-1986.

Though this paper concentrates mainly upon reinforced concrete structures as these are most readily repaired, consideration is also given to damage and repairability of other types of engineering structures and materials.

STRUCTURES

In general, the types of structures which are presently liable to damage from fire can be divided into groups according to the main construction materials used :

o   Reinforced concrete
o   Steel
o   Masonry
o   Timber
o   Other

There are, of course, a number of buildings which use other materials, such as aluminium, copper, glass, glass reinforced cement or plastic, but these tend to be a minority and the scope of this paper will not allow more than a passing reference to these materials.

The structural form of a building, a prestressed concrete frame for instance, may cause particular problems for assessment and repair and the modern type of steel composite decks may well present particular repair problems. However, in the majority of cases, the building will be a structure which is amenable to a relatively simple assessment and repair.

Timber and masonry form components of many types of structure but tend to be major structural components only in small domestic structures where major rebuilding is often the normal "repair" technique.

Steel structures, protected or exposed, tend to be repairable by re-use of undistorted existing members and replacement of all others with new materials and occasionally by reinforcing.

FIRE DAMAGE

Though fire damage is generally in the form of a local heat problem which may cause material damage, the whole of the fire experience must be considered, together with the age and use of the building and any other factors. If, for instance, the whole of the structure has aged or been damaged in some way such as surface carbonation of the concrete, then local repairs to fire damage are less sensible.

A fire experience may consist of heat causing local and, or, global damage, smoke damage to finishes or other absorbent surfaces, wetting by fire fighting with associated thermal shock effects and, in many cases, actual impact damage due to partial collapse.

Fire damage can be remote from the fire as, for instance, when floor expansion pushes walls over or when expansion causes strain damage at the base of columns or similar positions. A rise of 400°C would not in itself damage a structure but in a 10m length it could cause expansion of 50mm which could cause gross overstrain.

Penetration of water during or after a fire can also be a cause of damage and water inside a building will often penetrate remote areas by means of service ducts and may eventually manifest itself as severe condensation.

It is not possible to list all of the likely damage effects but the following is a brief summary of some of the more important ones :

- o   Concrete spalling
- o   Loss of strength of some steels
- o   Loss of strength of concrete
- o   Distortion of members
- o   Loss of strength of connections
- o   Moisture penetration
- o   Cracking of concrete
- o   Quenching damage
- o   Loss of heat treatment
- o   Smoke damage

ASSESSMENT

It is convenient to consider assessment in two ways, firstly as a Qualitative exercise, then as a Quantitative one. This approach allows for a more practical and cost effective approach to repairs as it tends to highlight the problem areas at an early stage.

The importance of the assessment of fire damage can not be over stressed as it is the key to any subsequent repair and its economic and practical viability. When carrying out an assessment, the individual concerned should be experienced in likely repair techniques and their limitations.

When developing a programme for assessment and repair, it is useful to consider a structured approach to the problem and the flow chart (Figure 1) gives an indication of how such an approach might be made.

QUALITATIVE ASSESSMENT

The general qualitative assessment method gives a simple way to indicate the degree of damage and likely repair of individual members and overall the structure. Figure 2 shows part of a first survey on a four-storey warehouse. The number adjacent to each member indicates the visual degree of damage according to the classification system given in Technical Report 15(1) which is to be standardized to make it simpler to use (Table 1). A detailed first assessment of this type (Figure 3) enables a very good initial picture to be drawn of the overall condition of the structure and the likely repair. At this stage, any surprising aspect in the performance of a structure would be noted.

The foregoing assessment procedure is for a reinforced concrete structure but a similar approach could be devised when considering a steel framed structure. A different damage classification table would be necessary as its main criteria would be based upon distortion of the steel. However, such an approach would allow for a good picture of the damage to be built up to assist in planning further assessment and repair.

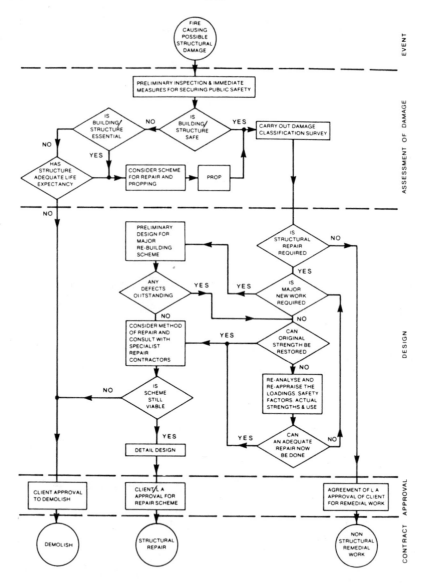

FIGURE 1: ASSESSMENT PROCEDURE FOR FIRE DAMAGED STRUCTURES

Table 1 Visual damage classification for reinforced concrete elements

Class	Element	Surface Appearance				Structural Appearance		
		Plaster/ Finish	Colour	Crazing	Spalling	Condition of main reinforcement	Cracks	Deflection
0	Any	Unaffected or beyond extent of fire						
1	Column	some peeling	normal	slight	minor	none exposed	none	none
	Floor	"	"	"	"	"	"	"
	Beam	"	"	"	minor	very minor exposure	"	"
2	Column	substantial loss	pink*	noticeable	localized to corners	up to 25%. none buckled	none	none
	Floor	"	"	"	localized to patches	up to 10%. all adhering	"	"
	Beam	"	"	"	localized to corners minor to soffit	up to 25%. none buckled	"	"
3	Column	total loss	buff/ friable	extensive	considerable to corners	up to 50%. not more than one buckled	minor	none
	Floor	"	"	"	considerable to soffit	up to 20%. generally adhering	small	not significant
	Beam	"	"	"	considerable to corners sides soffit	up to 50%. not more than one buckled	"	"

242

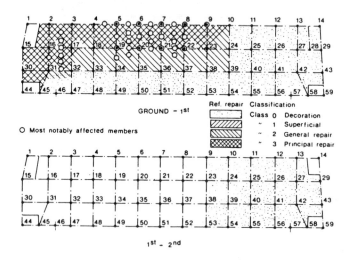

Ref. Classification factors – Table 1

C	3
B	2
A	1

4

Col

Member Ref

FIGURE 2: MEMBER CLASSIFICATION (GRD – 1st)

GROUND – 1st

O Most notably affected members

Ref. repair Classification

Class 0 Decoration
" 1 Superficial
" 2 General repair
" 3 Principal repair

1st – 2nd

FIGURE 3: INITIAL CLASSIFICATION OF REPAIR

QUANTITATIVE ASSESSMENT

The detailed quantitative investigation using various tests is to determine the likely temperatures to which the member has been subjected and hence the loss, if any, in strength. From examination of the building contents, it is possible to assess the maximum likely fire compartment temperatures and, from eye witness reports and evidence of combustible content, to assess the equivalent standard fire period. This can be compared with the temperature distribution and strength loss determined from test evidence. Report 15(1) indicated three main tests which could be carried out in order to determine strength loss in concrete. These are : colour determination, Schmidt hammer and core testing. The revised report deals with these and, in addition, gives guidance on the use and selection of other tests such as ultrasonic pulse velocity (5), Windsor probe (5), BRE internal fracture (5) and thermoluminescence tests (6). An example of tests carried out on a structure is shown in Figure 4. The information obtained from tests is then used to assess the strength of the member.

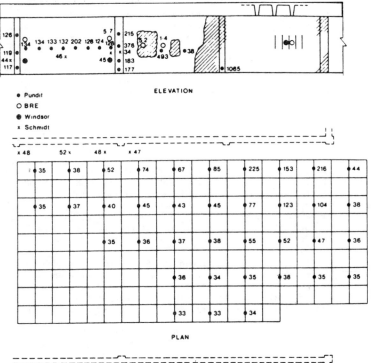

FIGURE 4 TESTS TO WALL AND WAFFLE

The strength reduction curves for concrete given in the previous report were based on research on unstressed concrete (7). It is known that the strength reduction of concrete unstressed at the time of heating is greater than for concrete under compression. Since the majority of

structural concrete will be stressed (at least under dead load) at the
time of heating, then some modification to the previous data is
appropriate. The strength loss data has been simplified to a straight
line plot for consistency with other documents (8,9)(Figure 5).
The colour of concrete can change as a result of heating (10) and therefore may be used to indicate the maximum temperature attained and the
equivalent fire duration. In some cases at above 300°C a pink
discoloration may be readily observed. The onset of noticeable pink
discoloration is important since it coincides approximately with the
onset of significant loss of strength due to heating. Therefore, any
pink discoloured concrete should be regarded as being suspect.

Pink discoloration is due to the presence of ferrous salts in the
aggregate and/or in the sand and, in some cases, these are not present.
Therefore, concrete which has not turned pink is not necessarily
undamaged by fire. The colour change to pink tends to be more
prominent with siliceous aggregates. Calcareous and igneous rushed rock
aggregates are less susceptible to this effect.

The residual strength of reinforcement can be judged from the
assessed temperature of the surrounding concrete or, if spalling has
occurred, from the period of exposure and likely fire compartment
temperature.

The effect upon steel reinforcement during elevated temperatures
and after subsequent cooling has been researched in detail (11,12).
Significant loss of strength may occur while the steel is at high
temperature and this is usually responsible for any excessive residual
deflections. However, recovery of yield strength after cooling is
generally complete for temperatures up to 450°C for cold worked steel
and 600°C for hot rolled steel. Above these temperatures, there will
be a loss in yield strength after cooling. The actual loss in strength
depends upon the heating conditions and type of steel but the simplified
values given in Figure 6 will be sufficient for most purposes. Where
this aspect is critical to the assessment, the matter should be discussed
with the reinforcement manufacturer if known or, alternatively, tests
carried out on samples taken from the member. Loss in ductility may
occur after exposure to particularly high temperatures. Buckling of
reinforcing bars often occurs as a result of compressive stress induced
at high temperatures by restraint against thermal expansion.

A damage factor may be determined for each member although normally
it will only be necessary to consider Class 3 and Class 4 damaged
members in detail. The damage factor is determined from Figure 5 using
the estimated average temperature within the compression block obtained
from temperature profile curves as, for example, are given in Figure 7.

The general process for this is to record the depth of the layer
which has reached a temperature of 300°C by reference, where appropriate,
to the depth of the pink zone. Estimate the increase beyond this depth
to reach a zone which has not exceeded 100°C by reference to Figure 7.
The average damage factor may then be determined for the compression
block taking a factor of 1.0 for all concrete subjected to temperatures
less than 100°C, and a factor of 0.85 for concrete within the 300°C to
100°C zone.

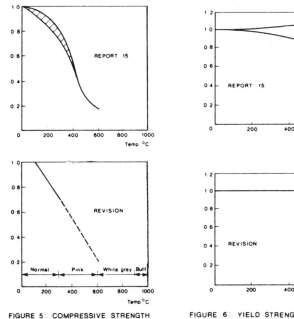

FIGURE 5 COMPRESSIVE STRENGTH
ON COOLING

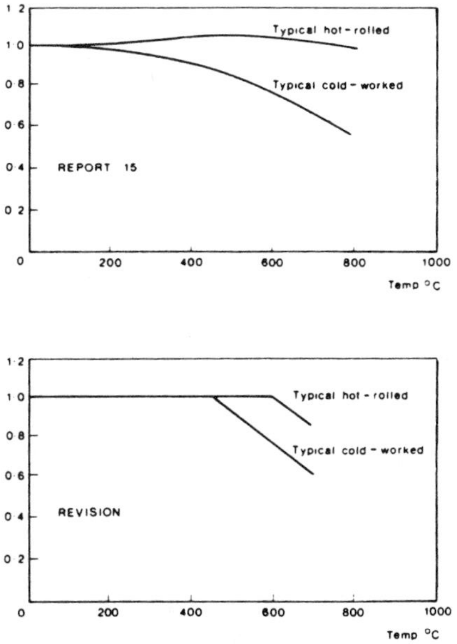

FIGURE 6 YIELD STRENGTH OF STEELS AT ROOM
TEMP AFTER HEATING TO AN ELEVATED TEMP

It may also be necessary to determine a damage factor to allow for
reduction in bond or anchorage of steel reinforcement. A number of
papers have been written (13,14,15,16). These indicate that bond is
influenced not only by temperature but also by bar diameter, type of
aggregate, section size and the compressive strength of the concrete.
The presence of links or other confining steel is also likely to
influence bond. A conservative damage factor for bond of 0.7 may be
applied to reinforcement within the 300°C to 100°C zone although a value
of 0.8 or better might be considered for lower temperatures, small
diameter bars (12 mm or less), concrete of lower compressive strength
(25-30 N/mm^2), concrete containing calcareous or lightweight aggregates
and for reinforcement contained by stirrups.

Damage to structural steel is generally confined to physical
distortion, as at about 550°C collapse will be imminent. Few steels are
permanently damaged by temperatures below 550°C and this subject is
covered by the paper by C.I. Smith and others, "The Reinstatement of Fire
Damaged Steel Structures" (17). The use of H.S.F.G. bolted connections
and many types of special bolts in connections, many of them being subject
to complex heat treatment, makes any such joint highly suspect and
specific tests would be necessary for any such joint where heat effects
could have been experienced.

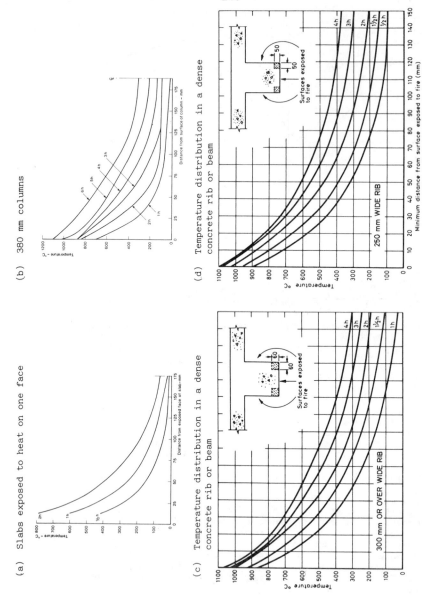

(a) Slabs exposed to heat on one face

(b) 380 mm columns

(c) Temperature distribution in a dense concrete rib or beam

(d) Temperature distribution in a dense concrete rib or beam

FIGURE 7 : Temperature gradients at various times from start of BS furnace test

Wrought iron and cast iron are also relatively unaffected by heat, up to collapse, though quenching effects on cast iron could be damaging.

Losses in strength of steel of up to 30% have been recorded (17) when temperatures in excess of 750°C have been reached and this could cause a local problem in some cases.

Testing of steel or iron is generally by hardness measurement or metallographic assessment and the test methods are given in the paper by Smith (17).

Damage to masonry can be occasioned and the normal effects are loss of strength of the mortar, in a similar fashion to normal concrete. Some mortars, particularly those using OPC, may lose strength rapidly after fires of a short duration but lime mortars are possibly less sensitive.

The writer is not aware of any document which would give guidance which would allow the assessment of a fire-damaged masonry wall in accordance with the current accepted design standards in UK. Repair must then be upon a replacement in full basis and the individual judgement of the Engineer must be exercised.

REPAIR TECHNIQUES

In addition to reviewing the sprayed concrete process (Gunite) which is the usual method of repairing major fire-damaged concrete structures, the working group has considered and included in the revision of Report 15, other repair methods such as resins, polymer modified and cement mortars, plaster, sprayed mineral preparations, as well as the use of alternative supports.

It is essential that a repair must restore any loss of strength, maintain durability and fire protection. In situations where, following a fire, there is still sufficient strength and cover for durability,then a thin hand or spray applied material could be used to restore a loss in fire protection. Equivalent thickness of concrete cover for various materials may be found from manufacturers. Some information is given in reference (22) and in BS 8110 (23) the revision to CP 110 (8).

Steel structures can be reinforced but the normal limiting factor is distortion and this will often require that replacement be implemented.

Resin Repairs

Resin repairs are commonly used to overcome problems of reinforcement corrosion but may not provide the necessary protection in the event of a subsequent fire.

Resin repairs may consist of a variety of configurations of patch or infill of Epoxy, Polyester and Acrylic mortar. Resins are often used for repairs to lightly spalled areas and, though they may perform quite satisfactorily in normal service, there is no comprehensive information on the performance of such repairs or that of the materials when subject to heat or an actual fire test. What information does exist, including some published papers (17,18,19), indicates that these materials may soften at relatively low temperatures (80°C). As a consequence, it is possible that some resin repairs may not provide adequate fire protection to the reinforcement and may fail to retain adequacy in compression zones. Accordingly it is recommended that resin repairs only be used when either:

(i)   performance data can be supplied to show that the particular formulation has adequate fire resistance and retains its structural properties under the envisaged fire condition, or

(ii)   the material is adequately fire protected by other materials and retains its structural properties at the expected fire temperatures at the relevant depth in the section, or

(iii)   loss of form of the material will not cause unacceptable loss of structural section or fire resistance.

The designer should refer to the manufacturers' literature for details of the performance of the various materials available. A wide variety of materials exists and their specifications are liable to changes at relatively short notice. In general, the materials will be capable of providing good bond and compressive strengths, the flexural and tensile strengths may exceed that of concrete but the thermal expansion is considerably larger than concrete and this may be a point to be considered where the temperature range is large. For further information, the designer should refer to the Concrete Society Technical Report No.26, Repair of concrete damaged by reinforcement corrosion (21).

Polymer modified mortars

In many cases, hand repairs of small areas can be effective by use of polymer-modified cementitious mortars. These repairs will generally be to areas or patches of between 12mm and 30mm depth. In particular, styrene butadiene rubber (SBR) modified mortars appear suitable. There is limited test information on such mortars but it is expected that they will be satisfactory in a fire as they should behave as a cementitious product. It is to be expected that the use of a small sized aggregate will improve the performance of mortars compared to concretes due to a lesser tendency for damage caused by aggregate splitting in the event of a subsequent fire. In other respects, these mortars are also described in the Concrete Society Technical Report No.26, Repair of concrete damaged by reinforcement corrosion (21) to which the designer should refer.

Cement mortars

These may be hand applied to damaged areas but great care in surface preparation is necessary in order to ensure adequate adhesion. Generally, mortars will be applicable to well-defined areas placed in layers using good rendering practice, up to a total 30mm thickness.

Plaster

This may be readily applied to both plain and roughened concrete surfaces. It can restore a degree of fire resistance but will not replace cover requirements for durability.

Sprayed mineral preparations

This technique will not assist when strength restoration or replacement of cover requirements for durability are required. Where internal repairs of a minor nature are necessary, these systems will restore fire resistance and shape to damaged members. For particulars, the designer should consult specialist contractors.

Alternative supports

The designer can consider the use of alternative supports such as further columns or new beams to sub-divide floor spans. Such schemes may well prove economic as they may allow lesser restoration to damaged members. New supports could be in reinforced concrete, steel, timber or masonry, in accordance with the normal new construction code of practice, to provide the required strength, fire resistance, durability and appearance.

ACKNOWLEDGEMENTS

This paper depends to a large extent upon the work of the Group revising the Concrete Society Report No.15 "Assessment of fire damaged concrete structures and repair by Gunite", and co-members of this group are :

Mr. A.K. Tovey   - C & CA (Chairman)
Mr. R.H.Jackson - Andrews Kent & Stone
Mr. E. Mellor   - De Leuw Chadwick O LEocha

REFERENCES

1. THE CONCRETE SOCIETY. Assessment of fire-damaged concrete structures and repair by gunite. Report of a Concrete Society Working Party.
London. The Concrete Society, 1978

2. GREEN, J.K. Technical Study: reinstatement of concrete structures after fire. The Architects' Journal. Vol.141, No.2, 14 July 1971. pp 93-99. No.3, 21 July 1971. pp 151-155.

3. SMITH, L.M. The assessment of fire damage to concrete structures. Thesis for PhD degree, Paisley College of Technology, Paisley, Scotland, September 1983. (Bibliography contains 259 relevant references)

4. TUCKER, D.M. and READ, R.E.H. Assessment of fire-damaged structures. Garston Building Research Establishment, November 1981 4pp. IP 24/81.

5. KEILLER, A.P. A preliminary investigation of test methods for assessment of strength of in situ concrete. Wrexham Springs, Cement and Concrete Association, 1981. 36pp. Technical Report 42.551.

6. PLACIDO, F. Thermoluminescence test for fire-damaged concrete. Magazine of Concrete Research. Vol.32, No.111. June 1980.

7. MALHOTRA, H.L. The effect of temperature on the compressive strength of concrete. Magazine of Concrete Research. Vol.8, No.23. August 1956. pp.85-94.

8. BRITISH STANDARDS INSTITUTION. CP 110: Part 1 : 1972. The structural use of concrete. Part 1 : Design, materials and workmanship. London. pp.156.

9. Design and detailing of concrete structures for fire resistance. Interim guidance by a Joint Committee of the Institution of Structural Engineers and The Concrete Society. London, The Institution of Structural Engineers, 1978.

10. BESSEY, G.E. Investigations on building fires. Part 2: The visible changes in concrete or mortar exposed to high temperatures. London. H.M. Stationery Office, 1950. pp.6-18. National Building Studies Technical Paper No.4.

11. STEVENS, R.F. Contribution to discussion on: Steel reinforcement, by R.I. Lancaster. Structural Concrete. Vol.3, No.4. July/August 1966. pp.184-185.

12. HOLMES, M., ANCHOR, R.D., COOK, G.M.E. AND CROOK, R.N. The effects of.elevated temperature on the strength properties of reinforcing and prestressing steels. The Structural Engineer. Vol. 60B, No.1. March 1982. pp.7-13.

13. MORLEY, P.D. and ROYLES, R. The influence of high temperatures on the bond in reinforced concrete. Fire Safety Journal, 2 (1979/80). pp. 243-255

14. SAGER, H. and ROSTASY, F.S. The effect of elevated temperatures on the bond behaviour of embedded reinforcing bars. Bond in Concrete: Proceedings of the International Conference, Paisley, 14-16 June 1982. London, Applied Science Publishers, 1982. pp.206-216.

15. ROYLES, R., MORLEY, P.D. and KHAN, M.R. The behaviour reinforced concrete at elevated temperatures with particular reference to bond strength. Bond in Concrete: Proceedings of the International Conference, Paisley, 14-16 June 1982. London, Applied Science Publishers, 1982. pp.217-228.

16. MORLEY, P.D. and ROYLES, R. Response of the bond in reinforced concrete to high temperatures. Magazine of Concrete Research. Vol.35, No.123. June 1983. pp.67-74.

17. C.I. SMITH, V.R. KIRBY, D.G. LAPWOOD, K.J.COLE,A.P.CUNNINGHAM and R.R. PRESTON. The reinstatement of fire damaged steel framed structures. Fire Safety Journal, 4 (1981) pp.21-62 Elsevier Sequoia, SA, Lausanne

18. LEVITT, M. The fire resistance of resin jointed concrete. Proceedings of Conference on Plastics in Building Structures(1965) Plastics Institute. Pergamon Press.1966. Paper 12.pp.77-81.

19. PLECNIK, J.M., BRESLER, B., CHAN,H.M.,PHAM,M. and CHOA,J. Epoxy repaired concrete walls under fire exposure. Proceedings of the American Society of Civil Engineers, Structural Division. Vol.108, No.ST8. August 1982. pp.1894-1908.

20. PLECNIK, J.M, BRESLER, B., CUNNINGHAM, J.D. and IDLING, B. Temperature effects on epoxy adhesives. Proceedings of the American Society of Civil Engineers, Structural Division. Vol.106. No.stl. January 1980. pp.99-113.

21. THE CONCRETE SOCIETY. Repair of concrete damaged by reinforcement corrosion. London, 1984. Technical Report No.26.

22. FIP/CEB Report on methods of assessment of the fire resistance of concrete structural members. Wrexham Springs, Slough. Federation Internationale de la Precontrainte. 1978.

23. BRITISH STANDARDS INSTITUTION. BS 8110:1985. The structural use of concrete. Part 2 : Recommendations for use in special circumstances. London.